AUST

PLACE
NAMES

For Charlotte, David and Katy.

AUSTRALIAN PLACE NAMES

BRIAN AND BARBARA
KENNEDY

HODDER AND STOUGHTON
SYDNEY AUCKLAND LONDON TORONTO

First published in 1989 by
Hodder and Stoughton (Australia) Pty Limited,
10–16 South Street, Rydalmere NSW 2116

Text copyright © Brian and Barbara Kennedy, 1988, 1992.

This edition published 1992

National Library of Australia
Cataloguing-in-Publication entry

Kennedy, Brian, 1937–
    Australian place names.

    ISBN 0 340 56676 0

    1. Names, Geographical – Australia.
    I. Kennedy, Barbara.   II. Title.

919.4'0014

Typeset in 9/9pt Plantin by Egan-Reid Ltd, NZ
Printed in Australia by The Book Printer

Introduction

Australia has over four million place names so that a book of this size can attempt to give derivations for no more than a tiny fraction of the best known towns, cities and locations. Most Australians live in suburbs of our capital cities and this book concentrates on these locations. Suburbs of the cities of Newcastle, Geelong and Wollongong are also given special attention.

Nearly three-quarters of Australian place names are of Aboriginal origin. They cause special problems to anybody interested in the derivation of Australian place names. Many of them were recorded by early surveyors who simply noted down the name given by the local Aborigines. When asked the name of a place on the bank of a river, an Aborigine might reply with the word for sand, place of water, or good hunting, depending on what he was thinking. For this reason many place names have several possible meanings. For example Mildura in Victoria could mean 'sore eyes' or 'red sand'.

The problem is made worse by the fact that the local tribes around our capital cities had largely died out by the second half of the 1800s and records of their languages are sketchy to say the least. Meanings given to Aboriginal place names today depend largely on folklore because until the 1930s no systematic research was undertaken into Aboriginal languages.

It is, however, interesting to note that some names in the same areas have syllables in common. For example, in Queensland, Indooroopilly means 'gully of leeches'; Mutdapilly, 'boggy gully'; Yeerongpilly, 'gully of sand'; and Jeeropilly, 'flying fox gully'. In Western Australia, many place names, such as Pallinup and Quindalup, end in 'up', which means 'the place of' or 'the camp of'.

The first European place names were those of the sea explorers. The Dauphin map linked to the possible Portuguese navigation along the east coast shows seven place names but none of them are now used. The Dutch explorers added some place names that still survive. They include Keer Weer (Turn Again) by Jansz in 1606. Captain Cook charted the east coast of Australia in 1770 naming such places as Port Stephens and Cape Hawke after admiralty officials and places like Bustard Bay and Cape Tribulation after incidents that occurred during the voyage.

The explorers by land also had their desperate moments as names like Mount Hopeless and Mount Disappointment still testify. However generally they preferred to call the places they came across after British officials who have long since been forgotten. Who would now recall the Colonial Secretary, George Murray, if Sturt had not called Australia's best known river after him?

It should of course have remained Hume's River after the first explorer to see it. But there is little justice in names.

Later settlers added their own names to the list. A glance through the 100 000-odd entries in the Australian Gazetteer reveals a desolate picture. Stoney Creeks, Sandy Creeks and Dead Horse Creeks abound in this, the world's driest continent. It is little wonder that the homesick Englishmen preferred to call their farms after places they had come from. These in turn became our modern suburbs which today are a reminder of the local pioneers as well as the places they came from.

There is some debate as to the longest place name in Australia. In the

Aug–Sept 1956 edition of *South Australian Motor,* a Royal Automobile Association magazine, advertising agency J. Walter Thompson inserted an advertisement for Shell Petroleum which mentioned the name, Cardivillawarracurracurracurrieappalarndoo. When asked by the South Australian Names Board for the source of the name, neither the company nor the agency could give any details.

Another long name with no official support is Warrawarrapiraliliullamalulacoupalunya, collected by John Flynn in the 1930s.

Of the officially recognised names, two of the longer ones are Lake Cadibarrawirracanna in South Australia at nineteen letters and Carleecardoocoverner Pool in the De Grey River area near Port Hedland in Western Australia at twenty letters. But the most likely contender is the Geographical Names Board of South Australia's entrant, Kerlatroaboorntallina Springs at twenty-one letters. It is located in the region of Mount Kingston, east of Peake Creek railway siding, and is recognised on mapping in South Australia.

Place name boards have been established in all states of Australia and the authors of this book are most grateful for help given by these organisations. Most of the place names given in this book, however, have been gleaned from local histories too numerous to mention by name.

Acknowledgements

A number of the entries in this book have been taken directly from A.W. Reed's *Place Names of Australia* by kind permission of his son, Mr John Reed.

Special thanks also to Stewart Cockburn for permission to print some 20 of the 4000 historically significant place names collected by his late father, Rodney Cockburn, who was the first Australian honoured by election to the English Place Name Society. Those seeking further information on South Australia are recommended to Rodney Cockburn's book, *South Australia: What's In A Name?* (Axiom Publications, Adelaide).

The authors would like to thank the following people and organizations:–
> Geelong Historical Records Centre.
> Newcastle City Council.
> Manly Municipal Council.
> Place Names Committee of the Northern Territory.
> The Geographical Names Board of NSW.
> Mr Ron McLeod, Survey Co-ordination Branch, Department of Conservation, Forest and Land of Victoria.
> Mr J.F. Morgan, Nomenclature Advisory Committee, Department of Lands and Surveys, Western Australia.
> Mr M. Sincock and Mr M. Medwell of the Department of Lands, South Australia.
> The Nomenclature Board of Tasmania.
> The Queensland Place Names Board.
> Jillea Carney, Australian Broadcasting Corporation.

A

Abbotsford *NSW* is a suburb of the Sydney municipality of Drummoyne. It is named from a village near Melrose Abbey on the Tweed River in southern Scotland where Sir Walter Scott was living when he wrote the Waverley novels. 'Abbotsford House', which gives the Drummoyne suburb its name, was built in 1878 and still stands. The land was purchased from Thomas Mort on 28 November 1876. The property was subdivided and offered for sale on 2 October 1905.

Abbotsford *V* is a suburb of Collingwood in the Melbourne metropolitan area. It derives its name from the estate of John Orr and comes originally from the Tweed River ford used by the Abbot of Melrose Abbey in Scotland.

ABC Range *SA* is said to have been named because of a theory that there are as many separate hills in the range as letters of the alphabet.

Aberfeldy *V*, a former goldfield north of Walhalla, is named after Aberfeldy in Scotland.

Aberfoyle Park *SA* is on land owned by Christian Sauerbrier in the 1850s. Sauerbrier was originally from Scotland. Aberfoyle is a locality name in the County of Perth, Scotland, and was possibly the home town of the Sauerbrier family. John Christian Sauerbrier, who inherited the majority of the property, had changed his name by deed poll to John Chris Aberfoyle in 1917 because of anti-German feeling at that time. His name was mistakenly assumed to be of German origin. The name was first used on a plan of subdivision in 1924.

Acton *ACT* was a name given to this locality by Lieutenant Arthur Jeffreys RN in 1843 after a town in Denbighshire.

Adaminaby *NSW* Aboriginal, meaning 'camping place', or 'place for resting'. The pastoral station established by Cosgrove and York in 1848 was named 'Adaminaby'. A town-site was formed in 1885 and given the name, Seymour, but was changed to the present form the following year. In 1957 the site was covered by the rising waters of Lake Eucumbene, formed by a large dam in the Snowy River Scheme. The present township is ten kilometres from the original site. Unusual names often provide a challenge to ingenuity. An apocryphal story relates how a German gave a mine to his wife, Ada, and said 'Ada's mine it be'.

Adamstown *NSW* in the Newcastle area was called Adam's Town after Thomas Adam who took up land there in 1869. It was proclaimed a municipality in 1886.

Adavale *Q* originally Ada's Veil, the name recalls an incident when Mrs E.J. Stevens, who was travelling with her husband to Tintinchella in 1870, lost her veil during the crossing of Blackwater Creek. When the railway reached the township, the name was changed to Adavale.

Adelaide *SA* is called after Queen Adelaide, the wife of William IV who requested that the Queen's name should be conferred upon the future capital. She left part of her library to the city. The main thoroughfare was called King William Street. The Aboriginal name was Tandanya, meaning 'the place of the red kangaroos'.

Adelaide Street Names When it came to naming the streets and squares marked on the first plan of Adelaide, the task was entrusted to a committee. After that, any person who subdivided land and established a new street had the privilege of naming that street.

The squares were named as follows: Victoria was called after Princes Victoria, who was then heir to the throne of England; Hindmarsh was named after Governor Hindmarsh, who was a member of the naming committee; Hurtle was chosen for James Hurtle Fisher, the resident commissioner and a member of the naming committee; Light honours William Light, the Surveyor-General; Whitmore was called after Woolryche Whitmore, one of the colonisation commissioners for South Australia.

Wakefield Street was named after Edward Gibbon Wakefield, the founder of South Australia; Pulteney Street honours Admiral Sir Pulteney Malcolm, a friend of Governor Hindmarsh; Rundle Mall is named after John Rundle, an original director of the South Australian Company; Morphett Street honours Sir John Morphett, president of the Legislative Council and a member of the street naming committee.

Adelaide River *NT* was discovered by L.R. Fitzmaurice and C. Keys of HMS *Beagle* in 1839, and named in honour of Queen Adelaide, who was dowager queen at the time, and consort of the late king, William IV.

Adelaide, Port *see* **Port Adelaide**

Adventure Bay *T* was named in 1773 by Tobias Furneaux, who commanded the *Adventure* on Cook's second voyage. He sheltered in the bay on 12 March 1773.

Ainslie *ACT* honours James Ainslie, the first overseer of 'Duntroon' station from 1825 to 1835.

Airds *NSW* was named in 1810 during Governor Macquarie's first visit to the Campbelltown area. He named Airds after his wife's family estate.

Aireys Inlet *V*, a holiday resort forty-seven kilometres south-west of Geelong, was named after J.M.C. Airey, a former lieutenant in the Royal Navy who took up a station called 'Anglohawk' or 'Eyrie' in 1842.

Airport West *V*, a suburb of Keilor on the outskirts of Melbourne, takes its name from its position near Tullamarine, Melbourne's international airport.

Alamein *V* is a suburb of Camberwell, in the Melbourne metropolitan area. It is named after the battle in North Africa during World War II which involved Australian troops.

Albacutya, Lake *see* **Lake Albacutya**

Albany *WA* was established by Major Edmund Lockyer in 1826 as the first settlement in Western Australia. Lockyer named the place Frederickstown after Frederick, Duke of York and Albany, brother of George IV. The name Albany was used in official documents from 1832.

Albert Park *SA* is a suburb of the Corporation of Woodville, a city in the Adelaide metropolitan area. Albert Park was auctioned by Townsend, Botting and Co. at the Port Adelaide Town Hall in October 1877. It was named in honour of Prince Albert, the consort of Queen Victoria. The name was bestowed by W.R. Cave.

Albert Park *V* is a suburb of South Melbourne. It is named after Queen Victoria's husband, Prince Albert.

Albert River *V* flows into Corner Inlet. It was discovered and named by W.A. Brodribb in 1841 after Prince Albert.

Alberton *V* is a township by Albert River. Its name was recommended by Surveyor Townsend in 1842 to honour Prince Albert.

Albion *V* is a suburb of Sunshine, a city on the outskirts of the Melbourne metropolitan area. It is the location of the Albion Quarrying Company. The name, Albion, refers to England.

Albion Park *NSW* is located on Samual Terry's grant alongside the Macquarie Rivulet. After Terry's death in 1838 the property was taken over by John Terry Hughes who owned the Albion Brewery which later became Toohey's

Brewery. Albion is an old name for Britain and was originally suggested as the name for Sydney.

Albury *NSW* was named after the village of Albury, in Surrey, England, when it was gazetted by T.S. Townsend in 1839. Albury became a municipality in 1859. It was proclaimed a city in 1946. In 1974 Albury was linked with Wodonga, its Victorian border-city neighbour, to become Albury-Wodonga, Australia's only national growth centre.

Aldinga *SA* is the final spelling of a word that has masqueraded as *Bgalti-ngga*, *Aulkingga* and *Aldingli*, the meaning of which no one has ever been able to establish authoritatively. Widely differing opinions suggest 'tree district', 'much water', 'battle or burial ground', and 'open, wide plain', according to Rodney Cockburn's *What's In A Name*.

Alexandra *V* is a township on the Goulburn River, twenty-four kilometres west of Lake Eildon. It was named after either Princess Alexandra, later wife of Edward VII, or Alexander McGregor, Alexander Don and Alexander Luckie who shared in the discovery of gold there in 1866. It was first known as Red Gate Diggings.

Alexandria *NSW*, a Sydney suburb, was named after Princess Alexandra, wife of the Prince of Wales who later became King Edward VII. The borough became independent from Waterloo in 1868.

Alexandrina, Lake *see* **Lake Alexandrina**

Alfred Cove *WA* takes its name from the sheltered cove which forms its northern boundary. The cove was named after Alfred Waylen, the original grantee of land which took in most of the present day localities of Myaree and Alfred Cove. Waylen arrived in the colony on the *Skerne* in January 1930 and acquired a 'Villa Grant' at Point Walter where he lived.

Alice Springs *NT* honours the wife of Charles Todd, who was in charge of construction of the Overland Telegraph. The sandy watercourse in which the Alice Springs are located was renamed the Todd River after Sir Charles Todd. The township was originally gazetted under the name of Stuart. The name was changed to Alice Springs in 1933.

Allambee *V* is a dairying locality near Yarragon, Gippsland. It is an Aboriginal word meaning 'quiet resting place'.

Allambie Heights *NSW* is derived from an Aboriginal word meaning 'a peaceful place'. The Allambie estate was sold by auction in 1918.

Allawah *NSW*, a suburb of the Sydney municipality of

Kogarah, comes from an Aboriginal word meaning 'remain here'.

Allenby Gardens *SA* is a suburb of Woodville, a corporate city in the Adelaide metropolitan area. Allenby Gardens was formed by the Public Trustee, Mr W. Wright, who named the suburb in 1921. He cut up part of Coombe's Estate near Croydon (Coombe and Samary Streets are named after Samuel and Mary Coombe). The name honours Field Marshall Lord Edmund H.H. Allenby, first Viscount Commander-in-Chief of the Palestine Campaign in World War I and afterwards High Commissioner for Egypt.

Alligator River *NT* was discovered by Captain Phillip P. King in May 1818 and given that name by him because of his impression that the crocodiles there were alligators. Tasman had referred to the Crocodile Islands off the coast in the 1600s.

Alphington *V* a suburb of Northcote in the Melbourne metropolitan area. It is named after the birthplace of William Manning in Devonshire, England.

Altona *V* is an industrial city about ten kilometres south-west of the centre of Melbourne. The name, Altona, appeared on a map for the first time in 1861. It was named by a German called Taegtow, who lived in Williamstown about this time, and who came from Altona in Germany. He believed, correctly, that there was coal in Altona and in 1881 he helped to form the Williamstown (Taegtow) Prospecting Co. The company went into liquidation in 1884. Later discoveries of coal were abandoned in 1931 when the State Government decided to work the open cut mine at Morwell instead. Altona became a city in 1968.

Alvie *V* is a township sixteen kilometres north-west of Colac. It was named after the birthplace of James MacPherson Grant, Lands Minister, near Inverness, Scotland.

Amadeus, Lake *see* **Lake Amadeus**

Amaroo *NSW* is called after the pastoral station of this name. Several translations have been given for the Aboriginal name. They include 'beautiful place', 'red mud' and 'rain'. It was believed to be the meeting place of the Macquarie and Lachlan tribes.

American River *SA* is on Kangaroo Island. It was so named by the first settlers there because an American whaler was wrecked there. The date of the wreck is not known, but among the first visitors to the island were whalers and runaway convicts from Tasmania, and articles published in

11

Tasmania in 1826 show the river was then already known by its present name. So the shipwreck may have been as long ago as ten years before 1826. The shipwrecked crew built a boat from local trees and the place where the boat was launched was still visible when the first colonists to South Australia arrived there in 1836.

Amstel *V* is a suburb of Waverley in the Melbourne metropolitan area. It is named after Van Amstel who had a market garden there.

Anakie *V* is a dairying district twenty-nine kilometres north of Geelong. Aboriginal in origin, the word, *anakie-yowang* means 'twin hills'.

Andamooka *SA* was the aboriginal name for a large waterhole discovered by Stuart. Today it is an opal mining settlement. It is located about 130 kilometres from Woomera.

Angle Vale *SA* is an old locality name which was gazetted as a suburb of Adelaide in 1983. The locality originally got its name because of the angle formed by Heaslip, Frado and Angle Vale roads.

Anglesea *V* is a coastal resort thirty-six kilometres south-west of Geelong. It was formerly called Swampy Creek, then Angelsea River. The town is named after a large island off the coast of Wales.

Anna Bay *NSW* was a derivation of Hannah Bay, and so named, according to tradition, in memory of a boat, the *Hannah*, which was alleged to have been wrecked there. Teramby School, Birubi Point Cemetery and a cone-shaped well standing on land owned in the late 1800s by a pioneer called William Eagleton, are three historical interest points in the area.

Annandale *NSW*, a suburb of Marrickville in Sydney, was the name of a large part of Petersham that was owned by Colonel Johnston. The original estate was in this municipality spreading back from Parramatta Road towards Petersham Railway Station. Annandale House was built in 1799 and remained in excellent preservation right up to the time of its demolition in 1910. An avenue of Norfolk Island pines stretched from the gateway opposite where Johnston Street enters Parramatta Road. Annan was the place in the south of Scotland where Colonel Johnston came from.

Annangrove *NSW* is a locality in the Baulkham Hills Shire on the north-western outskirts of Sydney. It derives its name from 'Annangrove House', a residence in the locality which originally belonged to a son of Colonel Johnston of Annandale.

Apollo Bay *V*, a holiday resort south-west of Geelong, takes its name from the bay which Captain Loutit named after his schooner. In its early days the settlement was known as Middleton. In 1874, this was changed to Kambruk, an Aboriginal word meaning 'sandy place'. In 1952, it was changed to Apollo Bay, which had been its popular name for many years.

Appin *NSW* was named by Governor Macquarie in 1811 after the small coastal village in Argyllshire, Scotland, which was the birthplace of Elizabeth Macquarie.

Applecross *WA*, a suburb of Melville, a city in the Perth metropolitan area, was named by Sir Alexander C. Matheson, third baronet of an old Scottish family, after his home village of Applecross in Scotland. He bought the land from the Western Australian Land Company in 1896.

Apsley Strait *NT* was discovered by Captain P.P. King on 21 May 1818 in the *Mermaid* and named after Apsley, the Earl of Bathurst and Secretary of State for the Colonies.

Arafura Sea is an expanse of water that separates the northern coast of Australia from New Guinea. The name is believed to be derived from the Portuguese word *alfours* or *arafuras*, meaning 'free men', probably because it was originally given to the inland tribes of the Aroe Islands, who remained isolated from the coastal settlements.

Aramac *Q* was named by the explorer, William Landsborough, in 1895 from a contraction of the name Robert Ramsay McKenzie, who was Colonial Treasurer and a friend of Landsborough. The explorer is said to have carved 'R.R. Mac' on a tree on the town-site.

Aranda *ACT* comes from the name of the Aboriginal tribe in Central Australia sometimes known as Arunta.

Arapiles *see* **Mount Arapiles**

Ararat *V*, a township ninety kilometres west of Ballarat, was named by its first settler Horatio Wills, who reached the area in 1841. After reaching the top of a 600-metre hill overlooking good pasture land, he wrote: 'This is Mount Ararat, for, like the Ark, we rested here'. Gold was discovered in May 1857.

Archipelago of the Recherche *WA* was visited by Admiral Joseph Antoine Raymond de Bruny D'Entrecasteaux in December 1792. It was named by D'Entrecasteaux after his ship *La Recherche* meaning 'search'. It was anglicised from 'L'Archipel de la Recherche' to 'Archipelago of the Recherche' by Matthew Flinders in January 1802.

Ardlethan *NSW* Meaning 'high' or 'hilly' in the Gaelic original, it was named after a place in Scotland.

Ardrossan *SA* was proclaimed in 1873 and was named by Governor Fergusson after Ardrossan in Ayrshire, Scotland, an electorate which he represented in Parliament after serving in the Crimean War. It contains the Gaelic roots *ard* 'height' and *ros* 'a prominent rock or headland'.

Ardtornish *SA* was formerly a locality in the Adelaide metropolitan area, but is now included in the suburbs of Holden Hill and Modbury in the corporate city of Tea Tree Gully. It was called after the earliest farm in the area, the 'Ardtornish Estate', which was established in 1839. The family that owned it called it after their native Ardtornish in Argyleshire, Scotland. The estate was managed by Miss Gregorson and her two nephews, Gillian and Angus Maclaine. Angus later gave half an acre of land to be set aside for school purposes. The school was subsequently converted into a woolshed and the estate was absorbed into the 'Beefacres' property. However, the gift was remembered when a new local primary school was built in the area and named Ardtornish.

Arkaba *SA* is a contraction of an Aboriginal name, Arkabatura, being the name of a tribe whose country was 100 kilometres north of Port Augusta. It was proclaimed on 18 January 1877.

Arkaroola *SA* in the northern Flinders Ranges is an Aboriginal word meaning 'the place of Arkaroo', a great legendary Dreamtime snake. It drank Lake Frome dry, carved out the Arkaroola Creek, and filled it by making water.

Armadale *V* is a suburb of Prahran, in the Melbourne metropolitan area. It gets its name from 'Armadale House' in Kooyong Road, built in 1876 by James Munro, a land speculator and Victorian premier, who called his house after a place in Inverness in Scotland.

Armadale *WA* is a town on the south-eastern outskirts of the Perth metropolitan area. The name was chosen by the railways department for its siding. It probably came directly from the place in Scotland, rather than Armidale in New South Wales, which is a corruption of the correct spelling of the name of the Scottish town.

Armidale *NSW* was named by G.J. McDonald, the Commissioner of Crown Lands. He called the settlement Armidale after his ancestral home in Scotland. Armidale experienced rapid growth. It became a town in 1849, a municipality in 1863, and a city in 1885.

Arncliffe *NSW* is a suburb of the Sydney municipality of Rockdale. It is named after a grant made in 1883 to David Hannan the first settler in the district. Hannan was Government Overseer of Brickmaking and the land surveyor, William Meadows Brownrigg, suggested that the grant be called 'Arncliffe' after a town in West Riding, Yorkshire, England. The name is shown as Arneclif in the Domesday Book of 1066 and means 'Eagle Cliff'.

Arnhem Land *NT* The name was given by Matthew Flinders in honour of the Dutch vessel that skirted the coast of the Northern Territory in 1623. In January of that year, Jan Carstensz in the yacht, *Pera*, accompanied by a smaller vessel, the *Arnhem* (or *Aernem*), followed the course taken by the *Duyfken* in 1606. The *Arnhem* became separated from the *Pera*, and discovered the north-east coast of Arnhem Land. Shortly afterwards, a landing was made on the coast of New Guinea where the master and ten of the crew were killed by natives.

Artarmon *NSW* is a suburb of the Municipalities of Willoughby and Lane Cove on Sydney's North Shore. Artarmon is said to get its name from the family estate in Ireland of William Gore, one of the early settlers and a former provost-marshal.

Arthur's Seat *V* is a mountain south-west of Dromana on the Mornington Peninsula. Arthur's Seat was named by Lieutenant Murray, who entered Port Phillip Bay in the brig *Nelson* in February 1802. He climbed Arthur's Seat, and called it after a similar mountain on the outskirts of Edinburgh, associated with Arthurian legends.

Arundel *V* is a suburb of Keilor on the outskirts of Melbourne. It is called after a place in Sussex, England.

Ascot Vale *V* is a suburb of Essendon, a city in the Melbourne metropolitan area. It is named after an English racecourse. Ascot is an appropriate name for a suburb so close to Flemington racecourse.

Ashburton *V* is a suburb of Camberwell, a city in the Melbourne metropolitan area. It was called after Ashburton Terrace in Cork at the suggestion of Councillor Dillon in 1890 when the railway station was named.

Ashburton River *WA* was discovered in 1861 by F.T. Gregory, and named by him in honour of Lord Ashburton, President of the Royal Geographical Society.

Ashbury *NSW* is a suburb of the Sydney municipality of Canterbury. Ashbury got its name because it was between

Ashfield and Canterbury.

Ashfield *NSW* is a municipality in the Sydney metropolitan area. By coincidence one of its early owners was the merchant Robert Campbell who was related to the Laird of Ashfield in Argyllshire, Scotland. But the name bestowed by Joseph Underwood who bought the land from Campbell in 1817 and named his new estate Ashfield Park after his birthplace in Ashfield, Suffolk, England. The name means 'field with ash trees'.

Ashford *SA* is a suburb within the corporate city of West Torrens in the Adelaide metropolitan area. Ashford recalls Dr Charles Everard, who acquired the property, 'Ashford Estate', soon after his arrival in the colony in 1836. He is remembered by an old gum tree behind Ashford House School, Anzac Highway, Ashford. He introduced the pomegranate to South Australia. There is another Ashford in Kent, England, from which the name is derived.

Aspendale *V* is a suburb of Chelsea, a city on the outskirts of the Melbourne metropolitan area. It developed around a racecourse that operated from 1893 to 1931. The name was taken from a successful racehorse named Aspen owned by J.R. Crooks and Whittington about 1882. The mare won two Newmarket handicaps.

Asquith *NSW*, a suburb in Hornsby Shire on the North Shore of Sydney, was named by the subdivider, Arthur Rickard, a well-known estate agent, after the British prime minister. Many of the streets are called after members of the British Cabinet.

Athelstone *SA* is a suburb of the corporate city of Campbelltown in the eastern metropolitan area of Adelaide. It takes its name from the property established and named by William and Charles Dinham, who bought land in the area in 1849. They erected the second flour mill in the colony. It was worked by water power. There is an Athelhampstone in Dorset, England.

Atherton *Q* is named after John Atherton, an early settler who arrived in the district in 1857.

Attack Creek *NT* On his courageous attempt to cross Australia from south to north, J. McDouall Stuart and his two companions were attacked here by Aboriginals. As the men were exhausted, the horses starving, and their provisions nearly all gone, the hostility of the Aboriginals convinced Stuart of the need to give up the attempt at that time. The date was 26 June 1860. The indomitable explorer vowed to return. Less than a year later he succeeded in returning to the creek (which he then named), and penetrated a further 160 kilometres north.

Attadale *WA* was originally part of land assigned to Archibald Butler in April 1830. In 1878 the block passed into the hands of John Butler of Milsons Point, Sydney, and was sold to A.P. Matheson eighteen years later. It was probably the latter who gave the area its present name. Matheson was a Scot and named nearby Applecross after a fishing village in his homeland. Attadale is also a town in Scotland and is situated on the eastern shores of Loch Carron, not far from Applecross.

Atwell *WA*, a suburb of Cockburn, a city south of Fremantle, was named in 1973 after Ernest Atwell, a Fremantle livery stable owner who bought the land now named after him in 1896.

Auburn *NSW* is a municipality located in the western part of the Sydney metropolitan area. Auburn was originally to be called Burford. Mr J.G. Mills wanted this name because his parents came from the Oxfordshire town of this name. The Railways Department objected as the name resembled Burwood and might have caused confusion. Mr Mills then decided upon the name Auburn after Oliver Goldsmith's poem, *The Deserted Village*.
   'Sweet Auburn, loveliest village of the plain
   Where health and plenty cheered the labouring swain'.

Audley *NSW* in Sutherland Shire, south of Sydney, was surveyed by George Edward Thickness-Touchet, 21st Baron Audley, in 1863-1864 where he set up a semi-permanent camp. He later became a son-in-law of Surveyor-General, Sir Thomas Mitchell.

Augusta, Port *see* **Port Augusta**

Augustus, Mount *see* **Mount Augustus**

Austinmer *NSW* is named after Mr Austin, one of the owners of the North Illawarra Coal Mining Company.

Austins Ferry *T* is named after James Austin, who began the first ferry service about 1818.

Austral *NSW*, a suburb of Liverpool, south-west of Sydney, was originally settled by a number of people including John Gurner, a solicitor, who arrived in the Colony in 1817. He requested his grant in 1826 and cleared and fenced it to run horses and cattle. Gurner Avenue is named after him.

Australia is a name that comes from the Latin word *australis*, meaning 'southern'. Perhaps its earliest form, Terra

Australis, meaning 'South Land', referred to a supposed southern continent. It was used on several early maps, notably in the world chart of 1569 prepared by the Flemish geographer, Gerhardus Mercator.

The Portuguese navigator, Pedro Fernandez de Quiros, reached the New Hebrides in 1606 and gave the name, Australia del Espiritu Santo, 'Southland of the Holy Spirit', to all of the southern regions as far as the South Pole. From 1642, when the Dutch explorer, Abel Janszoon Tasman sighted the coast of Tasmania, which he considered part of the southern continent, the continent became known as Nova Hollandia, 'New Holland'.

The earliest recorded use of the present form, Australia, appears to have been in Alexander Dalrymple's collection *Voyages of the South Seas* (1770). Captain James Cook referred to the continent as New Holland, but in 1770 claimed the eastern coast in the name of New South Wales. Captain Arthur Phillip's commission of 1786 constituted him Governor of New South Wales.

Matthew Flinders, in the preface to his *Voyage to Terra Australis* (1814), noted that 'Had I permitted myself any innovation upon the original term "Terra Australis" it would have been to convert it into Australia'. Governor Lachlan Macquarie began using the name, Australia, in official correspondence after 1817 when he received Admiralty charts based on the work of Flinders. These charts used the terms 'Australia' and 'Great Australian Bight'.

The British government continued to prefer the name, New South Wales, in its official correspondence. But Macquarie had given official approval to a word that was already gaining in popular use in the colony. By 1824, when the charts of Captain Phillip Parker King's maritime survey were published, the name, Australia, was in common use.

Australind *WA* was the site of an unsuccessful settlement between 1841 and 1845. It was intended that the town should be a centre for trade between Australia and India— hence the name.

Avalon *NSW* is a suburb of Warringah Shire in Sydney's northern beaches area. Avalon is a name which has links with King Arthur's mythical island and was chosen by the late Arthur J. Small when he subdivided the area in the 1920s. According to a remarkable story, Small woke up in the middle of the night shouting that Avalon would be the name of the subdivision. Unfortunately he did not leave any further explanation.

Avoca *NSW* is named after Avoca River in County Wicklow, Ireland. It was the 'sweet vale' of Thomas Moore.

Avoca *V* is a township twenty-six kilometres west of Maryborough. Major T.L. Mitchell named the river in 1836 after a stream in Ireland that was associated with the poet Thomas Moore.

Avon River *V* is the name of two rivers. The stream rising to the north-west of Avoca, which flows north-west to join the Richardson River, was named by Major T.L. Mitchell in 1836 from Shakespeare's Avon. The other is a river which rises near Mount Wellington in Gippsland and flows south-east into Lake Wellington. This river was named by Angus McMillan in 1840 after a river in Scotland.

Avondale Heights *V* is a locality in Keilor, a city on the outskirts of Melbourne. Avondale means 'valley by a river'.

Axedale *V* is a pastoral district twenty-one kilometres east of Bendigo. This and Axe Creek were named after the Axe River in Dorset, England.

Ayers Rock *NT* was visited by the explorer, William Gosse, in 1873 and named in honour of Sir Henry Ayers who was then Premier of South Australia. Its Aboriginal name, Uluru, has no literal translation into English.

Ayr *Q* was surveyed in 1882 and named by the Premier of Queensland, Sir Thomas McIlwraith, after his Scottish birthplace.

Bacchus Marsh *V* is named after Captain W.H. Bacchus, who made the district his home in 1838. The 'Manor House' he built there still stands. He is buried at the Holy Trinity Church.

Badgerys Creek *NSW*, a suburb of Liverpool, is named after James Badgery who came to New South Wales as one of the first free settlers in 1799.

Bairnsdale *V*, a town in East Gippsland, got its name in 1841 when a squatter, Archibald McLeod, called his station Bernisdale after his grandfather's village in Scotland. There is also a legend, for which there seems no real justification, that Bairnsdale was so called because of a great number of babies (bairns) born there in the early days of settlement.

Balaclava *V* is a locality in St Kilda in the inner Melbourne metropolitan area. It is named after the Russian battlefield in the Crimean War.

Balcatta *WA* is a suburb of Stirling, a city in the Perth metropolitan area. It is sometimes said to derive from the Aboriginal words, *bal* meaning 'his', and *katta* meaning 'hill', but this interpretation is questionable because the name was originally applied to a swamp and not a hill. The name was first recorded by Alexander Forrest in 1877 and was the Aboriginal name for the northern portion of Careniup Swamp. A later owner James Arbuckle, named his house 'Balcatta' after the area it overlooked and when the land around Careniup Swamp was subdivided it was given the name of Balcatta Estate. During World War I, North Beach was known locally as Balcatta Beach and in 1930 the road leading to the beach was gazetted as Balcatta Beach Road. This has since been changed to North Beach Road. The present Balcatta District boundaries were first officially defined in 1955.

Balcombe *V* is named after a pioneer, Alexander Balcombe, who lived in 'The Briars', a local historic mansion, from 1863 until his death in 1877.

Balga *WA*, a suburb of Stirling, a city in the Perth metropolitan area, is an Aboriginal name for the grass tree often known as black boy.

Balgowlah *NSW*, a suburb of Manly, is an Aboriginal word. Some books give the meaning as 'no devil', but this is pure conjecture. The Village of Balgowlah was named by Surveyor Larmer in 1832. It also appeared on old maps as Balgowla.

Balhannah *SA* was surveyed in 1839. It was laid out in 1840. James Thomson of Scotland was the founder and named the town after his mother, Belle, and sister, Hannah.

Ballam Park *V* is a locality in Frankston on the outskirts of Melbourne. The name recalls the estate once owned by Frederick Evelyn Liardet who built 'Ball Park House', now in Cranbourne Road, Frankston, in 1850. Ballam comes from an Aboriginal name meaning 'butterfly'.

Ballan *V* is a township thirty-eight kilometres east of Ballarat. It was named by squatter Robert von Steiglitz after his birthplace in Ireland.

Ballarat *V*, the state's largest inland city, takes its name from an Aboriginal word meaning either 'resting on one's elbow' or 'resting place'. The local city council insists that the name should be spelled Ballaarat but nobody else spells it that way. The Register of Place Names in Victoria gives it as plain old Ballarat. It was first called Yuille's Swamp. Gold was discovered there in 1851. The township was surveyed by W.A. Urquhart in 1852.

Ballendella *V* is a rural district eleven kilometres north-west of Rochester. It was named after an Aboriginal who accompanied Major T.L. Mitchell on his 1836 expedition.

Balliang *V* is a rural locality twenty-one kilometres south of Bacchus Marsh. It was named after an Aboriginal leader employed by Foster Fyans, a Crown Lands Commissioner in the 1840s, at his house at Marnock Vale, Geelong.

Ballina *NSW* is a corruption of an Aboriginal word, but it has been interpreted in various ways. James Ainsworth, who settled on the banks of the Richmond River in 1847, said that the name was Bullenah, which had some connection with fish or oysters. The town was gazetted in 1856.

Balls Head and **Balls Head Bay** *NSW*, localities in North Sydney, are named after Henry Lidgbird Ball, a naval officer who commanded HMS *Supply* in the First Fleet and who discovered Lord Howe Island. He became a pioneer explorer of Sydney's North Shore.

Balmain *NSW* gets its name from a grant made by Governor Hunter in 1800 to William Balmain, a surgeon in the First Fleet who had become Sydney's principal surgeon. Apart from conferring his name upon the district, Dr Balmain's connection with it was slight. In 1801, fifteen months after receiving the grant, he sold it for the token sum of five shillings to his friend John Gilchrist and returned to England. Balmain's main street is named after Governor Darling who was Governor of New South Wales from 1825 to 1831.

Balmattum *V* is a farming district ten kilometres north-east of Euroa. It is an Aboriginal word meaning 'man lying on his back', a description of the appearance of Mount Balmattum.

Balmoral *NSW*, a locality in Mosman on Sydney's North Shore, was named after the royal castle of Queen Victoria, built in 1853 at a cost of one hundred thousand pounds at Braemar, Aberdeenshire.

Balmoral *V* is a township near the Rocklands Reservoir, sixty-three kilometres north of Hamilton. It is named after the royal residence in Scotland. The district was originally called Mathers Creek or Black Swamp, being descriptive of a local feature. The Aboriginal name was Daarangurt.

Balonne River *Q* There are three possible meanings of this Aboriginal name: 'river'; *baloone* or *ballon*, said to mean 'pelican'; *baloon*, the local term for 'stone axe'. The river was discovered in April 1846 by Sir Thomas Mitchell, who used what he believed to be the Aboriginal name. One account, however, states that when he asked local Aboriginals for the name of the river he pointed to it with a stone tomahawk which he proposed to offer as a gift, and that in return he was given the word for a tomahawk. On the other hand it seems unlikely that he would be in the possession of such a primitive implement.

Balwyn *V* is a suburb of Camberwell, a city in the Melbourne metropolitan area. The name comes from Andrew Murray, a Scottish journalist with an interest in wine. He called his house 'Balwyn', home of the vine. He is said to have made up the word himself from the Gaelic *bal* and the Saxon *wyn*.

Banana *Q* is a shire centred on Biloela, about 100 kilometres south of Rockhampton. It is a rich farming, grazing and mining region, but there are no bananas! The name comes from a bullock named Banana which became so well-known that stockmen named the gully where it died, Banana's Gully. In 1858, gold was discovered in the gully.

Banksia *NSW*, a suburb of the Municipality of Rockdale, was named by the naturalist David Stead to honour Sir Joseph Banks.

Banksmeadow *NSW* is a suburb of the Sydney Municipality of Botany. It is named after Sir Joseph Banks. Captain Cook reported seeing 'as fine a meadow as ever was seen' somewhere on the shores of Botany Bay.

Bankstown *NSW* is a city in the western Sydney metropolitan area. Bankstown was named Banks' Town by Governor Hunter in honour of Sir Joseph Banks who had 'supported the colony in numerous ways'. Banks had accompanied Captain Cook on his voyage to Australia in 1770 and later he earned himself the title of 'Father of Australia' by urging the British Government to set up a colony at Botany Bay.

Bannockburn *V* is a township twenty kilometres north-west of Geelong. It was named after the Scottish town which was the scene of a famous battle in 1314.

Banyule *V* is a suburb of Heidelberg, a city in the Melbourne metropolitan area. It is named after one of the oldest surviving mansions in Victoria. It was built in 1842 for Joseph Hawdon who in 1838 became the first man to bring cattle overland from New South Wales to Port Phillip.

Bar Beach *NSW* in the Newcastle area got its unofficial name from a natural rock pool known as the Bar. The area was uninhabited in the 1920s and the young men from nearby Cooks Hill provided the personnel for the surf club which still has the name Cook's Hill Surf Club. However, when a new subdivision was established residents adopted the unofficial name Bar Beach for the locality.

Barcaldine *Q* in central Queensland was named after Barcaldine in Ayrshire, Scotland. The local pronunciation is Bar-CALL-din rather than BAR-call-deen.

Barcoo River *Q* is an inland river which flows to Cooper Creek. Edmund Kennedy found the river in 1847 and retained its original Aboriginal name.

Bardwell Valley and **Bardwell Park** *NSW*, suburbs of the Municipality of Rockdale, were named after Thomas Bardwell, an ex-convict who received a land grant about 1830.

Barellan *NSW* An Aboriginal word meaning 'meeting of the waters', or possibly, 'bowels'. It was the name of a local sheep station.

Bargo *NSW* A corruption of the Aboriginal name Barago, 'thick scrub', or 'brushwood'. It was sometimes called Bargo Brush.

Barker, Mount *see* **Mount Barker**

Barkly Tableland *NT, Q* This area of 140 000 square kilometres was discovered by William Landsborough in 1861 when in search of the Burke and Wills expedition and named by him after Sir Henry Barkly, who was then Governor of Victoria.

Barmera *SA* is an Aboriginal name which was once applied to Lake Bonney as well as to the settlement on its shore. It is a corruption of Barmeedjie, the tribal name of the local Aboriginals. *Barmera* is Aboriginal for 'water' or 'lake'. The town was proclaimed in 1921.

Barnawatha *V* is a farming district twenty-one kilometres west of Wodonga. It was originally called Indigo Creek. The Aboriginal name *barne-wathera* means 'deaf and dumb'.

Barongarook *V* is a farming district south of Colac. It is an Aboriginal word for 'running water'.

Barossa Valley *SA* takes its name from the Barossa Range. By a curious coincidence the range was named by Colonel William Light after Barrossa, a village in a wine-growing district in Spain, south-east of Cadiz, though in actual fact he chose the name because his friend, Thomas Graham (Lord Lyndoch), had served in the Peninsula War and won the battle fought at Barrossa in 1811. Barrossa means 'hillside of roses'. It was misspelt by a draftsman on an early map and the misspelling has now become the official Australian spelling. Light, who was South Australia's first surveyor-general, visited the district in 1837. A few years later the German geologist, Johann Menge, reported that the valley would prove a good locality for wine grapes. The Aboriginal name for the Barossa Peak is Yampoori, meaning 'little grass hill' because of the diminutive species of yacca growing there.

Barraba *NSW* The name of a station owned by J. Joskins in 1838. The Aboriginal name was Taengarrah Warrawarildi, 'place of yellow-jacket trees'.

Barrenjoey Peninsula *NSW* extends from Mona Vale to Palm Beach. Barrenjoey is derived from the Aboriginal word *joey* meaning 'young kangaroo'.

Barrington Tops *NSW* are named after Robert Barrington Dawson, son of the first agent of the Australian Agricultural Company which first settled the Gloucester district in the 1890s.

Barron River *Q* was named by the police inspectors R.A. Johnstone and A. Douglas in 1876 in honour of their friend T.H. Barron who was a clerk in the Queensland police department. The river had been discovered a year earlier by J.V. Mulligan, who was under the impression that it was the Mitchell River.

Barrow Island *WA* was discovered by P.P. King in June 1818 and named after Sir John Barrow, who was Under-Secretary to the Admiralty in 1818. The north-west portion of this island was originally sighted by a French expedition in 1803 who named several capes but presumed it to be part of the mainland.

Bartle Frere, Mount *see* **Mount Bartle Frere**

Barton *ACT* honours Sir Edmund Barton (1849-1920) who became the first prime minister of Australia in 1901.

Barwon River *V* was named by Surveyor Wedge, who called the river Kondak Baarwon, meaning 'great', 'wide' or 'deep water'.

Basin, The *see* **The Basin**

Basket Range *SA* has two recorded sources to the derivation of the name:
　　1. German emigrants 'squatted' in the vicinity of Basket Range. The women had very large baskets which they filled with produce and carried on their heads to markets.
　　2. The splitters in The Tiers, as the Mount Lofty hills were once called, had to pay a licence fee. Mr Basket had control of that department and collected the fees. His hut stood at the foot of the hill, just below Ashton Post Office, and was called Basket Bottom.

Bass *V* is a township by Bass River north-east of San Remo. It was named after Surgeon George Bass, who discovered Western Port.

Bass Hill *NSW* is a suburb of Bankstown, a city located in the western part of the Sydney metropolitan area. Bass Hill recalls George Bass who first explored the Georges River with Matthew Flinders and received a grant of land on Prospect Creek in 1798.

Bass Strait *T* was named after George Bass, who, with his friend Matthew Flinders, circumnavigated Tasmania in 1798 and proved it was an island.

Bassendean *WA* is a town in the Perth metropolitan area. Peter Broun, who later became the first colonial secretary of Western Australia, selected a grant on the west side of the river on 29 September 1829. He built a residence to settle his wife and young family, and named it 'Bassendean' after his family seat in Berwickshire in Scotland.

Batemans Bay *NSW* originally appeared on Captain Cook's chart as Bateman Bay, being discovered on 21 April 1770 and named after Nathaniel Bateman, Captain of the *Northumberland*, on which Cook had sailed as Master.

Bathurst *NSW* was named after Henry, 3rd Earl Bathurst (1762-1834), Secretary for the Colonies 1812 to 1827. The name was given by G.W. Evans, who reached the plains to the west of the Great Dividing Range in 1813, and set up his camp on the present site of Bathurst. When Governor Macquarie traversed the road built over the Blue Mountains some eighteen months later, he recorded the name of the future town officially as Bathurst. Gold was discovered near Bathurst in 1851. It became a city in 1885.

Bathurst Island *NT* In 1819, P.P. King proved that Bathurst and Melville Islands were separated from each other. They had been seen in 1644 by Tasman, who thought they were a part of the mainland. King named the smaller one after Earl Bathurst.

Batlow *NSW* was the name of a pastoral holding when the area was first surveyed by Mr Townsend in 1853.

Batman *V* is a locality in the City of Coburg in the Melbourne metropolitan area. The name honours John Batman, the father of Melbourne, who landed at the present site of the city in May 1835.

Batman's Hill *V* honours John Batman, the founder of Melbourne, who arrived at Port Phillip in April 1836 and lived in a small house on this hill. The hill was removed in the

1860s to provide a site for railway yards and goods sheds.

Battery Point *T* is a locality name which followed from the establishment in the 1820s of the Mulgrave Battery on the point overlooking Sullivans Cove. Princes Park includes the site of the battery.

Baulkham Hills *NSW* is a shire on the north-western outskirts of the Sydney metropolitan area. Baulkham Hills, according to the historian James Jervis, got its name from Baulkham Hills in the county of Roxburgh in Scotland where Andrew McDougall, one of the local pioneers, came from.

Bayswater *V* is a suburb of Knox City on the outskirts of the Melbourne metropolitan area. The name comes from the birthplace of John James Miller, the first shire president. His homestead at Boronia was named Bayswater and it became the name of the district after 1895.

Bayswater *WA* is a shire within the Perth metropolitan area, and one of the city's inner suburbs. The shire got its name from the London suburb of the same name because the land belonging to a Mr Baynard was crossed by a creek called Baynard's Water and in time this was corrupted to Bayswater. Sale of property here known as 'Bayswater' was reported in *The Morning Herald* newspaper on 31 July 1885. Bayswater was included as a stopping place on the Eastern Railway in a timetable gazetted on 31 March 1892.

Beachport *SA* was founded in 1878. Sir Samuel Way named it in honour of Sir M.E. Hicks-Beach, a secretary of state for the colonies who, on his elevation to the peerage, adopted the title Earl of St Aldwyn. According to Rodney Cockburn, the Booandik tribe of Aboriginals called it Wirmalngrang, 'owl's cave'.

Beacon Hill *NSW*, a locality in Warringah Shire, was named by the Lands Department in 1881. The department gave it this name because a beacon or trigonometrical point had been established at the top of the hill. However, a great deal of much more colourful but inaccurate folklore has grown around the name. According to one story, beacons were lit to send smoke signals to Parrmatta when scouts sighted ships bringing fresh provisions to the starving colony, in the early days of the settlement.

Beaconsfield *T* was named by F.A. Weld, Governor of Tasmania, in honour of the Earl of Beaconsfield, the British statesman, Benjamin Disraeli. When Lieutenant Colonel Paterson established a settlement at Port Dalrymple, the site of Beaconsfield was known as Cabbage Tree Hill. When gold was discovered, it was renamed Brandy Creek, and did not receive its present name until 1879.

Beaconsfield *V* is a suburb of Berwick, a city on the outskirts of the Melbourne metropolitan area. It is named after Benjamin Disraeli, Earl of Beaconsfield who died in 1881 on the day a deputation called on the Victorian Minister of Railway, asking for a station to be built at this place.

Bealiba *V* is a rural locality near Cochrane, twenty-one kilometres north-west of Dunolly. It comes from an Aboriginal word *beal-ba* meaning 'flooded red gum creek'.

Beardy Plains; Beardy River; Beardy Waters *NSW* Colonel Dumaresq, the owner of 'Tilbuster' station some eight kilometres from Armidale, employed two stockmen named Duval and Chandler in the 1830s. They both had an intimate knowledge of the region round about, and were notable for their long, black beards. Newcomers to the region who were searching for suitable locations to establish their stations were usually advised to consult the 'beardies'. The nickname was eventually given to the district, and also to the two rivers named Beardy River and Beardy Waters. There is also a Beardy Street in Armidale. In more serious mood, Mount Duval Chandler River and Chandler's Peak provide a permanent memorial to these two notable stockmen.

Beaudesert *Q* is a town about sixty-five kilometres south of Brisbane. 'Beaudesert' station was named in about 1842 by Edward Hawkins, after a station in New South Wales which was in turn named after Beau Desert Park in Staffordshire, England.

Beaufort *V*, a township forty-seven kilometres west of Ballarat, was named after Rear Admiral Beaufort. It was first known as Fiery Creek.

Beaumaris *V* is a suburb of Sandringham in the Melbourne metropolitan area. It was named after the Welsh coastal resort where Edward I built Beaumaris Castle. Its former name was Spring Grove.

Beaumont *SA* was named by Sir Samuel Davenport (1818-1906), a pioneer legislator with cabinet rank, with influence in primary production circles. It means 'beautiful mountain'.

Beauty Point *NSW*, a locality in Mosman on Sydney's North Shore, was the name given to the headland when an estate there was subdivided. It used to be called Billygoat Point.

Beeac *V* is a township by Lake Beeac, nineteen kilometres north of Colac. It was named after a local Aboriginal tribal leader.

Beechworth *V* is a township and tourist resort thirty-six kilometres east of Wangaratta. Gold was discovered in 1852 and the area was originally named May Day Hills. It was surveyed in 1853 by Smythe who named the settlement after his birthplace in Leicestershire, England. The bushranger Ned Kelly was sent here from Melbourne by special train and preliminary proceedings in connection with his trial were heard in the local court.

Beecroft *NSW*, a locality in Hornsby Shire on the North Shore of Sydney, was part of the Field of Mars Common until 1874, when it was resumed by act of Parliament as crown land, to be used for houses. In 1886, when the northern railway line was opened, the district was surveyed and subdivided. Sir Henry Copeland, who was then Minister for Lands, gave the name Beecroft to the area. It was his wife's maiden name. Street names also recall Copeland's interest in the area. Apart from Copeland Road, there are Mary Street and Hannah Street, called after his first and second wives. Hull and Malton Roads were named after the places in Yorkshire, where he and his wives were born.

Beenleigh *Q* is the main centre of Albert Shire, which lies between Brisbane and the Gold Coast. The Beenleigh area was known to the Aboriginals as Woobbummarjo, meaning 'boggy clay'. Captain Logan, commander of the Moreton Bay penal settlement, explored the area in 1826. Governor Darling later named the Logan River in his honour. John Davey and Frank Gooding established the first permanent settlement about 1865. They began to grow sugarcane on their plantation, which was named 'Beenleigh' in memory of a family estate in Devonshire, England.

Bega *NSW* comes from an Aboriginal word meaning 'beautiful' or 'large'. On an early plan, the name appears as 'Bika'. It appeared as 'Biggah' in 1839, when the first licence was taken out for the station. By 1848, the licence was held by Dr George Finlay in the form of 'Bega', which was finally adopted for the town.

Belair *SA* is South Australia's first national park. It is located in the City of Mitcham in the southern metropolitan area of Adelaide. Old Government House in the park was built in the 1860s as a place for the governor to get away from the summer heat of Adelaide. *Belair* is French for 'beautiful air'. According to Rodney Cockburn and H.C. Talbot, there are two theories about the origin of the name. According to one version it was named by Gustave Adolph Ludewig after the birthplace of his wife in the French colony of Martinique. Another derivation of the name is that Eugene Bellairs, a government surveyor, lived in the area in 1849.

Belconnen *ACT* has been associated with the locality since the days of the early settlers.

Belfield *NSW* is a suburb of the Sydney municipality of Canterbury. It got its name because it is between Belmore and Enfield. The name first appeared on post office records when the Belfield branch of the ALP wrote asking the department to establish a post office at North Belmore.

Belgrave *V* is a locality in the Shire of Sherbrooke on the outskirts of the Melbourne metropolitan area. Its name comes from the name of a house, 'Mount Belgrave', built by William G. Benson. He took up land in 1879 and his holding included the whole of present day Belgrave. He called his estate after Belgrave in Leicestershire.

Bell *NSW* is a locality in the Blue Mountains west of Sydney. It was originally called Mount Wilson, when the railway platform was first constructed in the nineteenth century. However, because the town of Mount Wilson was many kilometres from the railway the name Bell was applied to honour Archibald Bell (1804-1883) who surveyed a road later to be known as Bell's Line of Road, linking Bell and Kurrajong.

Bell Bay *T* is believed to have been named after Charles Napier Bell, a noted marine engineer who surveyed many of the harbours and inlets around the coast of Tasmania between 1891 and 1914.

Bell Park *V* is a residential district in the Shire of Corio. It is probably named after Bell Post Hill. In 1903, John Bell, owner of Woolbrook station, bought Lunan estate by Geelong's Western Beach.

Bell Post Hill *V* is a residential district in the Shire of Corio. The hill is located just west of Geelong where squatters John Cowie and David Stead, who arrived in March 1836, erected a bell to warn station hands of any danger threatening from Aborigines.

Bellarine *V*, a rural locality near Geelong, comes from an Aboriginal word meaning 'elbow' or 'resting on one's elbow'.

Bellbird *NSW* Probably the name of the original coalmine.

Bellenden Ker, Mount *see* **Mount Bellenden Ker**

Bellerive *T* was initially called Kangaroo Point probably because of the numbers of kangaroos seen by the first settlers. The present name was given about 1832 and is derived from the French for 'beautiful river'.

Bellevue Hill *NSW* is located in the Municipality of Woollahra in the Sydney metropolitan area. Bellevue Hill was first called Vinegar Hill. (Not to be confused with the hill of the same name in Blacktown.) The name offended Governor Macquarie who in 1820 ordered 'a finger board to be painted with the name, Belle Vue, and fixed on the centre of a circle or mound on top of the hill vulgarly called Vinegar Hill'.

Belmont *NSW*, a locality near Newcastle, was named by Thomas Williamson who settled there. He called it after his home town, a village on the Island of Unst in the Shetland Islands.

Belmont *V* is a suburb of Geelong. The town was proclaimed within the district of Newtown and Chilwell (Geelong) on 15 April 1861. Dr Alexander Thompson squatted there in 1836. The name comes from Lancashire, England, but is of French or Italian origin meaning 'fine hill'. Belmont near Derry, Ireland, is the location of St Columb's stone.

Belmont *WA* is on land allocated in 1831 to Captain Francis Henry Byrne. It is believed that it was named by him after his estate in England. As early as August 1831, official records refer to the location as Belmont.

Belmore *NSW* is a suburb of the Sydney municipality of Canterbury. Belmore is one of the oldest settlements in the municipality. A school called 'Belmore' was established in 1869. It was named after the Earl of Belmore, Governor of New South Wales.

Belrose *NSW* is a suburb of Warringah Shire. Belrose is named after a combination of the flowers, Christmas bells and native rose.

Beltana *SA* comes from an Aboriginal word and probably means 'running water'. In 1855, 'Beltana' was the name of a sheep station owned by I. Haimes. The town of Beltana was proclaimed on 2 October 1873.

Ben Buckler *NSW* is now the site of a tower marking the outlet of the Bondi sewer, which was opened in 1889. Various people have suggested different origins for the name. One is that Governor Macquarie named it after the Scottish island Benbecula, but this is a small sandy islet, bearing no resemblance to the Bondi headland. The name may commemorate Ben Buckley, brother of William Buckley, the man who lived with the Aboriginals, and is said to have given rise to the expression 'Buckley's Chance'. But the most likely contender is an early settler named Ben Buckler, who was killed by the collapse of a shelf of rock on which he was standing in the vicinity of Bondi.

Ben Lomond *NSW* is named after the Scottish mountain. The New England town is on the western slope of the mountain that bears the same name.

Ben Lomond *T* was first sighted by Flinders. The mountain was later named by Colonel William Paterson, doubtless after the Scottish mountain, when he founded the first settlement in northern Tasmania in 1804. The Aboriginal name was Toorbunna.

Benalla *V* is a town south-west of Wangaratta on the Broken River. Its Aboriginal name comes from either *benalta* meaning 'musk duck' or *benalla* meaning 'large waterhole'. It was first called Broken River.

Benambra *V* is a rural township twenty-two kilometres north-east of Omeo. It is an Aboriginal name meaning 'hills with big trees'.

Bendemere *Q* is a shire located on the Darling Downs. The name appears to be derived from Bendemere holdings, taken up by Joseph King, whose tender for Bendemere and Inglebogie was accepted by the crown in 1854. The headquarters of the shire is the town of Yuleba.

Bendigo *V,* a city in central Victoria, is named indirectly after British pugilist Abednego William Thompson who was famous at the time the settlement was christened in 1851. The Gold Commissioner renamed it Sandhurst in 1853 because his father had been Governor of Sandhurst Military College in England. It was officially named Bendigo in 1891.

Bengworden *V* is a rural locality twenty-five kilometres south-west of Bairnsdale. It comes from an Aboriginal word for 'second'. It was once called Little Limerick because of the number of Irish descendants in the district.

Bennelong Point *NSW* The site of the Sydney Opera House commemorates one of the first Aboriginals to be introduced to European society. As early as 1789 Governor Phillip captured a young man named Bennelong, who gradually became accustomed to the food and clothing of his white captors. The Governor gave him a house on the eastern point of Sydney Cove (once known as Cattle Point), on which his name was later conferred. He lived there with his wife Barangaroo. His life was an uneasy one, halfway between two cultures. In 1792 he accompanied Governor Phillip to England and was presented to King George III. On his return his second wife Gooroobarrooboollo rejected him. It is thought that he was born in 1764 and was killed in a tribal fight in 1813.

Bentleigh *V* is a suburb of Moorabbin, a city in the Melbourne metropolitan area. It was named in honour of Sir Thomas Bent, a premier of Victoria and a local councillor.

Bentley *WA* got its name in 1940 when the Canning Road Board proposed the name Bentley Park because an area within the proposed district had been known locally as Bentley Hill for over seventy years. Enquiries as to the origin of the name indicated that there was a considerable camp established on this hill when the Albany 'Block' Road was being constructed about 1870, and that the warder in charge of the men was named Bentley. The name was approved by the Nomenclature Advisory Committee in May 1940 when the Committee adopted the general policy of deleting the suffix where applicable. e.g. Morley Park became Morley, but Victoria Park retained the Park to avoid confusion with the State.

Bermagui *NSW* An Aboriginal word meaning 'canoe with paddles'. It has also been spelt 'Bermaguee', while an early plan showed it as 'Permageua'.

Berowra *NSW*, a locality in Hornsby Shire on the North Shore of Sydney, is an Aboriginal word meaning 'place of many winds'.

Berri *SA* was surveyed by E.A. Loveday in 1910 and proclaimed and named by Governor Bosanquet in 1911. According to an article in the *Chronicle* of 15 February 1962, the name *berri berri* was Aboriginal for 'wide bend in the river'. The article says the name was shortened to Berri. However, the South Australian Geographical Place Names Board has no verification of this in its official records. The first recorded reference to this name was 'Beri Beri Hut', as shown on early survey records. The irrigation area and the town of Berri were probably named after the hut, which in turn probably derived from the Aboriginal word for a bush growing in the vicinity.

Berrima *NSW* is an Aboriginal word meaning 'southward'.

Berriwillock *V* is a rural township twenty kilometres south-east of Sea Lake. It comes from an Aboriginal word meaning 'birds eating berries'.

Berry Island *NSW* on the northern shores of Sydney Harbour, is named after Alexander Berry, a pioneer of North Sydney. It is no longer an island. Berry himself had a causeway built and the mud flats were finally filled in during the 1960s.

Berrys Bay *NSW*, a locality in North Sydney, is named after Alexander Berry (1781-1873), a Scottish merchant, who went into partnership with Edward Wollstonecraft.

Berwick *V* is a city about forty kilometres to the south-east of Melbourne. The name probably came from Berwick-on-Tweed in England, which was the home of both Captain Robert Gardiner, the first settler, and Robert Bain, founder of Berwick's Border Hotel. Berwick was created a city in 1973. The Old English word *berewic* means 'a demesne farm'.

Bessiebelle *V* is a rural locality thirty-two kilometres east of Heywood. It was named by Surveyor J.G. Wilmot, after Bessie Cameron.

Bethanga *V* is a rural township north-west of Tallangatta. The station was taken up in 1846 by R. and D. Johnston. Gold was discovered in 1853. *Bethanga* is an Aboriginal word for 'weed'.

Beulah *V* is a rural township thirty-five kilometres north of Warracknabeal on the Yarriambiack Creek. The name has a biblical derivation.

Beverley *SA* is a suburb of Woodville, a corporate city in the Adelaide metropolitan area. It was sold in 1849 by E.N. Emmett and Co. on 'very long credit if required', promising that good water was available at moderate depth. The name is from Yorkshire and means 'the beaver's haunt'.

Beverly Hills *NSW*, a suburb of the Sydney municipalities of Hurstville and Canterbury, was originally called Dumbleton after a farm at the corner of King George's and Stoney Creek Roads. When the railway line was opened on 21 December 1931, more people settled in the area. The newcomers considered the name of Dumbleton to be unattractive and likely to retard the progress of the district. The local organisations made representations to Hurstville Council and secured their support for a change of name. The name of Beverly Hills was chosen as it was one of the beauty spots in America. The name of the railway station was changed on 24 August 1940, and the post office on 1 October 1940.

Bexley *NSW*, a suburb of the Municipalities of Rockdale and Canterbury, is named after a land grant made to James Chandler in 1822. It is named after the village of Kent in which he was born.

Bexley North *NSW* is a suburb of the Sydney municipality of Canterbury. It was the name given to the railway station on the East Hills line, opened in 1931. It was on the 'Bexley Estate', the name given to his farm by James Chandler, after Bexley Heath in Kent. 'Bexley House' was located near the intersection of Bexley Road and Homer Street.

Bicheno *T* is named after James Ebenezer Bicheno, a popular colonial secretary of Van Diemens Land from 1843 until he died in Hobart in 1851.

Bicton *WA*, a suburb of Melville in the Perth metropolitan area, was the property of John Hole Duffield who arrived in Perth in 1831 and produced some of the first wine in the colony on his vineyard where the suburb of Bicton is now located.

Bidwill *NSW* is a locality in the City of Blacktown on the western outskirts of Sydney. It is named after John Carne Bidwill (1815-1853) who was the first director of the Sydney Botanic Gardens and Government Botanist.

Bilgola *NSW* is a beach in Warringah Shire. Bilgola comes from an Aboriginal word. Some books suggest that it means 'swirling water', but this is just conjecture. Surveyor James Meehan made the first attempt to spell the name Belagoula, in his survey of 1815. The present spelling comes from 'Bilgola House', built on the beach in the 1870s by W. Bede Dalley, a member of parliament.

Biloela *Q* comes from an Aboriginal word meaning 'cockatoo'. It is pronounced Bill-oh-wheel-a.

Binalong *NSW* Two contradictory suggestions have been offered—that it is an Aboriginal word meaning 'towards a high place' or that it was the name of a well-known Aboriginal named Bennelong who came from another part of the country. (It has not been suggested that this was Bennelong of Bennelong Point, q.v.) The same authority says that the local residents objected to Bennelong because it was not euphonious, and that they preferred to change it to Binalong.

Binda *NSW* An Aboriginal name meaning 'deep water'.

Bindi *V* is a farming locality twenty-five kilometres north-east of Swifts Creek. The Aboriginal word *bindi* means 'stomach'.

Binginwarri *V* is a rural locality north of Welshpool. The Aboriginal word means 'hungry stomach'.

Binna Burra *Q* is a popular resort in the Lamington National Park area. Binna Burra Lodge is named after the Aboriginal word for the white beech tree that grows in the park's rainforests. The Lodge was founded in 1933 by Romeo Lahey, a local sawmiller and early conservationist, and Arthur Groom, an author and explorer. Their aim was to enable visitors to stay and appreciate the beauty of the area.

Birchgrove *NSW*, the northernmost finger of Balmain, was a 1796 grant to a private in the New South Wales Corps, George Whitfield, who established an orange orchard there.

In 1810 it was sold to John Birch, paymaster to Governor Macquarie's newly-arrived 73rd Regiment. The new owner renamed the orchard 'Birch Grove' and built 'Birch Grove House', the first residence in the region. When it was demolished in 1969, it was one of the oldest houses in Australia.

Birchip *V* is a township north-west of Charlton. Settlement began in 1882 and the town site was surveyed in 1887 by surveyor Vernon. It is an abbreviation of the Aboriginal name Wirrembirchip meaning 'the ear'.

Birdsville *Q* was so named because of the plentiful bird life in the surrounding district.

Birregurra *V* is a rural township east of Colac by the Barwon River. It is an Aboriginal word meaning 'kangaroo camp'. It was known as Buntingdale, the site of the first Aboriginal mission in Victoria, and also as Bowden's Point.

Birrong *NSW* is a suburb of Bankstown, a city located in the western part of the Sydney metropolitan area. *Birrong* is an Aboriginal word meaning 'star'.

Bischoff, Mount *see* **Mount Bischoff**

Bittern *V* is a locality near Hastings. It was named after the nocturnal bird which haunts reed beds and swamps.

Black Forest *SA* is a locality in Unley in the Adelaide metropolitan area. The 'Black Forest' of trees was known by this name as early as 1836. It was the haunt of bushrangers and cattle thieves in the early days.

Black Rock *V* is a suburb of the City of Sandringham in the Melbourne metropolitan area. Black Rock is named after 'Black Rock House' in Ebden Ave, Black Rock. It was built in 1856 for Charles Ebden an, eccentric early settler, land speculator and an early member of the Victorian parliament. He named it after the birthplace in Ireland of his wife Tamar. The house is modelled on Belmont Castle near Cape Town. For a time it was the summer residence of the Governor of Victoria.

Black Stump *NSW Q V*, is the name of four locations. Two of them—one at Merriwagga, a village of seventy people between Griffith and Hillston about 600 kilometres west of Sydney, and Coolah, in north-west New South Wales—have rival claims to have been the original black stump of the expression 'this side of the black stump'. There are also black stump landmarks at Mundubbera in Queensland and Johnsonville, Victoria.

The Merriwagga claimants went to the trouble of erecting a black stump replica in 1970. Locals claim references to Black Stump tank, creek and swamp date back to the 1880s. According to local legend, the name had its origin in the tragic death of Barbara Blaine, a teamster's wife who camped at a waterhole sixteen kilometres south of the village in 1886. While her husband went off to find better pasture for his bullocks she lit a fire to prepare a hot meal. It was a hot, windy and dusty day and the grass caught fire. The teamster returned to find his wife burnt to death. He later described her as looking 'just like a black stump'.

On 13 March 1886, Mrs Blaine became the first person buried in Gunbar Cemetery near the once vast station of that name. The waterhole where she was camping became known as Black Stump Camp and the name was later passed on to the new tank nearby. The area is known as the Black Stump District and Telecom has officially recognised it as such with a telephone exchange listing under that name.

Coolah Shire has published a pamphlet setting forth its own claims to the black stump:

'There is evidence that several wayside inns existed in the Coolah Shire area. These were needed for travellers along the long road routes. The best known was the Black Stump Wine Saloon situated near the Gunnedah road, six miles north of Coolah. It was the staging post for north-western New South Wales. The importance of this inn resulted from its position at the junction of the old coach-road. It was also a resting place on the old Sydney stock route before passengers entered the rough country on the last leg of the journey.

'The position of the inn is clearly marked on old New South Wales Lands Department maps and like many a pub before and since, it became the hub from which men dated their journeys and gauged their distances. . .The saloon was named after the nearby Black Stump Run and Black Stump Creek, both of which derived their names from the local saying of "beyond the Black Stump".

'In 1826 Governor Darling proclaimed "limits of location" or boundaries "beyond which land was neither sold nor let" nor "settlers allowed". This boundary was located in 1829 as being the northern side of the Manning River up to its source in the Mount Royal Range, then by that range and the Liverpool Range westerly to the source of the Coolaburragundy River, then along the approximate location of the Black Stump Run, then in a south-westerly direction to Wellington.

'Land north of this location was referred to as land "beyond" and the use of the word "beyond" can be found in the *Government Gazette* of 19 January, 1837.

'However, settlers did not strictly adhere to the Governor's proclaimed boundaries and often let their stock graze "beyond". Thus in the Coolah area, to avoid detection by officialdom, the location of these pastures was vaguely described as being "beyond the Black Stump". . .

'It is understood that the Black Stump Wine Saloon was erected in the 1860s. The centre was then a small settlement possessing its own racetrack. The saloon was destroyed by fire in 1908.'

Blackall *Q* was first explored by Sir Thomas Mitchell in 1846. It was surveyed in 1868 and named after Sir Samuel Wensley Blackall, the second governor of Queensland.

Blackburn *V* is a suburb of Nunawading, a city on the outskirts of the Melbourne metropolitan area. It is said to have been called after a man named Blackburn who owned sawpits beside the creek which was given his name. Alternatively it may have been named after a place north-east of Liverpool in England.

Blackett *NSW* is a locality in the City of Blacktown on the western outskirts of Sydney. It is named after George Forster Blackett, superintendent of the Government Cattle Station at Rooty Hill (1820-1830).

Blackheath *NSW* is a locality in the Blue Mountains west of Sydney. It was named by Governor Macquarie during his tour over the Blue Mountains in 1816. He seems to have named the area because of the heath that grew there, rather than after the London suburb.

Blacktown *NSW* is a city on the western outskirts of Sydney. Blacktown owes its name to a school for Aboriginals which was established by Governor Macquarie at Parramatta under William Shelley. This school was later moved to the site where the Richmond Road meets Rooty Hill Road North. It was closed in 1824 under Governor Brisbane, but reopened from 1827 to 1833. An Aboriginal reserve was established on the opposite side of the road and the name Black's Town came to describe the area. Later the new road from Prospect to Richmond was identified as the Blacktown Road. In the 1850s the Railways Department adopted the name Blacktown for the station situated at the junction of the railway line and the Blacktown Road.

Blackwater *Q* is a coalmining centre in central Queensland, 190 kilometres west of Rockhampton on the Capricorn Highway. The township was laid out in 1886 and the name was inspired by the dark colour of local waterholes.

Blackwood *V* is a locality north of Ballan. It was named after Captain Francis Blackwood, who made the hydrographic survey of Australia's north-east coast and part of the New Guinea coastline.

Blackwood, Mount *see* **Mount Blackwood**

Blair Athol *Q* is the site of a small but economically important coal basin. James McLaren, who took up a run in the district in 1863, named his head station 'Blair Athol' after the seat of the Duke of Athol in Scotland. He discovered coal in 1864, when sinking a well for water.

Blair Athol *SA* is a suburb in the corporate city of Enfield in the Adelaide metropolitan area. The suburb takes its name from the home of the Magarey family. Thomas Magarey bought the land in the area in 1857-8. He was born in Ireland but lived in Lancashire, England, as a boy. Blair Athol Castle is in Perth, Scotland.

Blakehurst *NSW*, a suburb of the Municipality of Kogarah, was called after a pioneer named Blake.

Blaxland *NSW*, is a locality in the Blue Mountains west of Sydney. It was named after Gregory Blaxland (1778-1853), one of the three explorers who crossed the Blue Mountains in 1813.

Blayney *NSW* dated from 1843 when a design for a village church, at Kings Plains on a portion of the Church and school reserve to be called Blayney, was approved by the Governor.

Blinman *Q* After 'Pegleg' Robert Blinman, a shepherd with a wooden leg who was employed on the 'Angorichina' run. According to legend his favourite resting place was on a hilltop. One day, somewhere about 1859, he broke off a piece of rock and found traces of copper, and from this chance discovery came the rich peacock ore which led to the establishment of the coppermining town that was surveyed in1859.

Blue Lake *SA* is situated in the steep-sided depression of the old volcanic cone of Mount Gambier. The water in the old volcano is fresh. It is used as the town water supply for Mount Gambier. The colour changes from grey in winter to blue in summer. The reason for this change is not fully understood. The most acceptable theory suggests that the calcium carbonate dissolved in the water precipitates—comes out of the water as a solid—and the particles cause light scattering to occur. The lake is about twenty metres above sea level and about eighty metres deep.

According to Rodney Cockburn's *What's In A Name*, the lake was originally designated Lake Power, after Mrs David Power, wife of the largest land owners in the South-East. Another title was Devil's Ink Bottle. The Aboriginal name was Waawor. Governor and Lady Blanche MacDonnell were said to be the first persons who were rowed over the Blue Lake, which was thereupon rechristened Lake Blanche, but the new name did not endure. The lake was first reported by Tyers and Hoddle in 1840, although the first white man to see it was Stephen George Henty in 1839. It was surveyed and sounded in 1851 by Blandowski, who surveyed several pastoral leases in the region.

Blue Mountains *NSW* is a city sixty kilometres west of Sydney. The Blue Mountains were originally called Carmarthen Hills and Lansdowne Hills by Governor Phillip in 1788. Soon after, however, the name Blue Mountains was adopted because of the blue haze that enveloped the area. All distant hills look blue due to scattered rays of light in the intervening atmosphere. The blue light, which is of short wave-length, is scattered by contact with fine dust particles and droplets of floating water vapour. However, minute droplets of oil in the air can scatter blue light more effectively and the vast number of gum trees in the area emitting eucalypt oil makes the Blue Mountains look much bluer than most other distant hills.

Blues Point *NSW*, a locality in North Sydney, is named after Billy Blue, a part-Jamaican negro who became Sydney's first ferryman. He was sentenced to seven years transportation at the Kent Quarter Sessions in 1796 and arrived in Sydney in 1801. He became a boatsman on Sydney Harbour. In 1805 he married Elizabeth Williams and they had six children.

Bobbin Head *NSW*, a suburb of Ku-ring-gai Municipality, is named after a farm occupied by a Mr Hutchinson. It is said to come from an Aboriginal word meaning 'place of smoke'.

Bobs Farm *NSW* was named after Bob, a convict servant who acted as stockman to a land-holder called 'Gentleman Smith'. After long admiring the property he wished to acquire when he was 'out of his time', Bob died before his ambition was achieved. Eventually, Magnus Cromarty bought Smith's herd and became owner of the land coveted by the stockman, and ever after referred to it as 'Bob's Farm'.

Bogan Gate *NSW* Aboriginal. The name Bogan has several times been translated 'birthplace of a king', or 'birthplace of a great king'. As any concept of royalty was foreign to Aboriginal people, it would seem that the name is a tribute to an Aboriginal with extraordinary capacity for leadership.

Bogan River *NSW* The name has the same origin as Bogan Gate in the preceding entry. On 1 Januray 1829 it was named New Year's Creek, but this name, given by Charles Sturt when he discovered it, did not persist. In December 1828 Captain Sturt and Hamilton Hume had begun to explore the western course of the Macquarie. Their progress was blocked by swamps which led them to another creek which they named after the day of its discovery.

Boggabilla *NSW* An Aboriginal name meaning 'great chief born here'.

Boggabri *NSW* An Aboriginal name meaning 'place of creeks'. Other forms of the Aboriginal word that have been recorded are: Bukkiberai, Bukkibera, and Bukkabri. An unusual and less likely theory is that the word was Boorgaburrie, meaning 'emu', or 'young'.

Bogong *V* takes its name from the Bogong moth which was a delicacy favoured by the Aboriginals who used to gather the dormant larvae and cook them.

Bogong, Mount *see* **Mount Bogong**

Boisdale *V* is a rural township eleven kilometres north of Maffra by the Avon River. Its name is of Scottish derivation.

Bold Park *WA* commemorates Perth's town clerk W.E. Bold, who retired in 1944 after a record forty-four-year term of office.

Bolivar *SA* was named after the General Bolivar Hotel, a hotel built in the area by Mr Walpole who arrived on the sailing ship *Bolivar*. General Bolivar was known as the liberator of South America.

Bombala *NSW* An Aboriginal name meaning 'meeting of the water'.

Bondi *NSW* Sydney's most famous beach is located in the Municipality of Waverley. Bondi is an Aboriginal word which is said to mean 'the sound of waves breaking on the beach'. The first mention of the word is in the field book of the pioneer surveyor, James Meehan, who referred to it as Bundi Bay.

Bondi Junction *NSW* was called Tea Gardens in its earlier days but became Bondi Junction because the Waverley tram junction was located there.

Bonegilla *V* is a rural area twelve kilometres east of Wodonga. It is an Aboriginal word meaning 'deep waterhole'.

Bong Bong *NSW* is an Aboriginal word meaning 'out of sight', as the Wingecarribbee River seems to disappear in this area in dry weather.

Bongaree *Q* is the main township of Bribie Island, seventy kilometres north of Brisbane. Bongaree was a young Aboriginal who accompanied Flinders to Moreton Bay.

Bonnet Bay *NSW* in Sutherland Shire was named by the Geographical Names Board, New South Wales Lands Department, in 1969. An area of land north-west of Jannali and bounded by the Woronora River, it is named after a bay in the river. It is possible that the bay was named after a cave in the area which was known as The Old Woman's Bonnet—its shape being like an old-fashioned bonnet.

Bonney, Lake *see* **Lake Bonney**

Bonnie Doon *V*, a township by Brankeet Creek, north of Lake Eildon, was named by Thomas Nixon, after the famous stream in Scotland.

Bonnyrigg *NSW* is a suburb of Cabramatta on the south-western outskirts of the Sydney metropolitan area. It has a complicated derivation. The word *rigg* is used in the south and north-east section of Scotland and means 'a ridge' or 'land elevation'. The word *bonny* is also a Scottish expression and means 'pleasant' or 'pretty', so that Bonnyrigg denotes a picturesque district. The original village of Bonnyrigg in Scotland is located about six kilometres west of Falkirk, near the Grand Canal which links Gransmouth and Glasgow.

Booby Island *Q* On 23 August 1770 Captain Cook wrote in his Journal: 'Being now near the Island, and having but little wind, Mr Banks and I landed upon it, and found it to be mostly a barren rock frequented by Birds, such as Boobies, a few of which we shot, and occasioned my giving it the name of Booby Island.' Apparently unaware of this, Captain Bligh later recorded in his Journal: 'A small island was seen bearing W. and found it was only a rock where Boobies resort, for which reason it is called Booby Island.'

Boolaroo *NSW* is an Aboriginal word for 'two' or 'place of many flies'.

Boonah *Q* comes from an Aboriginal word, *buna*, meaning 'bloodwood tree'. It was first known as Blumbergville, after an early storekeeper.

Booragoon *WA* was the Aboriginal name for the lower reaches of the Canning River and was chosen as a locality name in the early 1950s by the then Melville Road Board.

Booragul *NSW* is an Aboriginal word meaning 'sunshine' or 'summer'.

Boorhaman *V* is a rural area north of Wangaratta by the Ovens River. The name is supposedly an Indian word for 'a ragged beggar'.

Boort *V* is a rural township fifty kilometres south-west of Kerang by Lake Boort. It is an Aboriginal word meaning 'smoke'.

Boosey *V*, a farming district west of Yarrawonga, comes from an Aboriginal word meaning 'gum tree'.

Bordertown *SA* is not, as its name suggests, on the border, but is situated 20 kilometres inside South Australia on the main interstate highway, 290 kilometres from Adelaide and 456 kilometres from Melbourne. Inspector Tolmer used 'Scott's Woolshed' in the vicinity of Border Town as a depot for gold escorts travelling from Mount Alexander in Victoria to Adelaide. It was changed from Border Town to Bordertown in the *Government Gazette* on 5 April 1979.

Boronia *V* is a locality in the City of Knox on the outskirts of the Melbourne metropolitan area. It was so named by A.E. Chandler in 1915 because of the sweet-smelling boronia shrubs growing in the nursery he had established in the district in 1895.

Bossley Park *NSW* is a suburb of Fairfield on the south-western outskirts of the Sydney metropolitan area. It is named after the Bossley family who pioneered the district.

Botany *NSW*, a Sydney municipality, takes its name originally from Botany Bay which was named by Captain Cook in honour of the new plants discovered there by the botanist, Sir Joseph Banks, in 1770. Botany was a village when the first municipalities were set up around Sydney in the 1850s. The three Botany municipalities—Botany, North Botany and West Botany—did not come into existence until the Municipalities Act of 1867. In 1886, the Government put forward a proposal for a new borough to be called Cooksdale after Captain Cook. The local residents were opposed to both the borough and the proposed name. In 1887, modified plans for two boroughs, Botany and North Botany, overcame the residents' objections and the Municipality of Botany was proclaimed in 1888. J.J. Macfadyen became the first mayor of the new borough. Later, North Botany became Mascot and West Botany became Rockdale. The municipality of Botany was then left to perpetuate the original name of the region. Botany and Mascot were finally amalgamated under the present name and boundaries in 1948.

Botany Bay *NSW* was originally called Stingray Harbour or Stingray Bay by Captain Cook. On 28 April 1770, Cook wrote in his journal: 'At daylight in the morning we discovered a Bay which appeared to be tolerably well sheltered from all winds, into which I resolved to go with the Ship, and with this in view sent the Master in the Pinnace to sound the entrance.' On 6 May he wrote: 'In the evening the Yawl return'd from fishing, having caught two sting rays weighing near 600 pounds. The great quantity of New Plants & ca. Mr Banks and Dr Solander found in this place occasioned my giving it the name of Botany Bay.'

Bothwell *T* was named after the Scottish town because most of the settlers came from Scotland.

Boulder *WA* was named after Great Boulder Mine, one of the first and richest mines in the Golden Mile. The town was proclaimed on 4 December 1896.

Bourke *NSW* was named in 1835 when the explorer, Thomas Mitchell, built a stockade eleven kilometres from the site of the present town. He named it Fort Bourke. A memorial cairn now marks Mitchell's original site, along with a replica of the original stockade. Bourke became a town in 1860.

Bowen *Q* was selected in 1859 by Captain H.D. Sinclair of the cutter, *Santa Barbara*, when he was searching for a port for the development of the Kennedy district. Port Denison was established two years later by G.E. Dalrymple, who named Bowen after the first governor of Queensland, Sir George Ferguson Bowen.

Bowen Hills *Q* is a locality in the Brisbane metropolitan area. It honours Sir George Ferguson Bowen, who became Queensland's first governor in 1859.

Bowenvale *V* is a rural locality ten kilometres north of Maryborough. It was named after Sir George Ferguson Bowen, Governor of Victoria from 1873–1879. It was formerly known as Timor.

Bowral *NSW* is an Aboriginal word meaning 'high and large'.

Box Hill *V* is a city located about fifteen kilometres from the centre of Melbourne. Its name was chosen by Silas Padgham, the first postmaster and storekeeper, in 1861 when the first post office was opened. It was named after Padgham's birthplace, Dorking, in Surrey, which is at the foot of Box Hill in England.

Boydtown *NSW* is an old whaling settlement south of Eden. Boydtown today consists of a hotel called the Seahorse Inn, which has relics of the whaling era in Twofold Bay, and the ruins of an old church. The inn was constructed by convict labour for Ben Boyd, the colourful adventurer, financier and whaler, who founded Boydtown in 1843. Boyd's idea was to make Boydtown a great commercial centre, a port to ship wool from Boyd's properties in the Monaro district and the most magnificent city in Australia. The dream did not come true. By 1847 Boyd's ventures were losing money, and his shareholders forced him out. Boyd left Australia in 1849. He was last heard of in 1851 on Guadalcanal in the Solomon Islands, where it is presumed he was killed by the islanders.

Bradbury *NSW* dates from a farewell visit that Governor Macquarie paid to Campbelltown in January 1822. He noted William Bradbury building a two-storeyed brick house for an inn on his farm and noted that Campbelltown was a good staging post for travellers. When Macquarie was invited by Bradbury to name the farm, he bestowed on it the title Bradbury Park. Bradbury had a tannery at the corner of Old Menangle Road and the Appin Road area, where the old stone cottage stands today.

Braddon *ACT* honours Sir Edward Braddon, one of the founders of the Constitution.

Bradleys Head *NSW*, a locality in Mosman on Sydney's North Shore, is named after William Bradley, a lieutenant on HMS *Sirius*, the flagship of the First Fleet. The Aboriginal name is Burrogy.

Brahma Lodge *SA* is taken from the trotting stud conducted on the land by Mr Frank Reiss. When he sold his land for subdivision in 1960, the name of his stud was retained.

Branxholme *V* is a rural township twenty-five kilometres south-west of Hamilton. It was originally named Tommy Best's Ford, and was renamed by Surveyor Lindsay Clarke.

Branxton *NSW* takes its name after a place in Northumberland, England.

Braybrook *V* was the original name of the locality where the City of Sunshine now stands. It was named after Braybrook in Berkshire, England.

Brays Bay *NSW* is a locality in the Sydney municipality of Concord. It is named in memory of John Bray who lived there.

Breadalbane *NSW* was named by Governor Macquarie in 1820 after a range of mountains sixty kilometres west of Perth in Scotland.

Breakfast Creek *Q* is a locality in the Brisbane metropolitan area and was named because John Oxley's party had breakfast there on 17 September 1824.

Breakfast Point *NSW* is a locality in the Sydney municipality of Concord. It was named by Captain John Hunter during his voyage up the Parramata River in 1788. The Aboriginals at the point shook hands in a friendly fashion and Hunter invited them to share his breakfast. The Aboriginals were not impressed with the white men's food but were happy to cook their mussels on Hunter's fire.

Breakwater *V* is a locality south of Geelong where a stone breakwater was built across the Barwon River.

Bredbo *NSW* was the name of a river and two pastoral holdings occupied by Patrick Clifford in 1866. It is probably an Aboriginal word.

Bremer Bay *WA* was named by John Septimus Roe after James Bremer, captain of the *Tamar*, under whom Roe served. The town-site (named after the bay on which it is situated) was declared on 20 July 1962.

Bremer River *Q* was named by the explorer John Oxley in 1824. He called it after Captain James Gordon Bremer, who was then in charge of HMS *Tamar*.

Brewarrina *NSW* is said to be an Aboriginal word for 'fisheries' because the Aboriginals built stone dams as fish traps in the river there. The town grew up in the 1860s after it became a crossing place for stock.

Brigolong *V* is a rural township thirty-four kilometres north of Sale. It is an Aboriginal word meaning 'a resort for native companions'.

Bribie Island *Q* is named after a convict.

Brickfield Hill *NSW* is a locality in the City of Sydney. It was the source of clay for the city's first bricks. The name survives today as a post office with that name.

Bridgewater *SA* is a name with a complicated origin. During the 1840s, the Government built a new road to the Mount Barker district. It by-passed the village of Cox's Creek and it was not long before the hotel there was moved to its present site on the new road. In 1855, when the licence was held by a Mr Addison, its name was changed to the Bridgewater Hotel. It is not known why he chose the name. Today it is one of the oldest hotels outside the Adelaide metropolitan area. In 1859 John Dunn laid out and named the township of Bridgewater. Dunn began a new industry in 1860 when he built the Bridgewater Mill. Bridgewater is also a private subdivision at Port Adelaide, which now forms part of the suburb of Birkenhead.

Bridgewater *T* is of uncertain origin, but it was in use by the mid-1930s, before a bridge was built over the River Derwent. The first bridge was opened in 1849. It is possible that it was named after Bridgewater in Somerset, with which the children of the Tasmanian town exchanged flags around the turn of the century.

Bridgewater *V* is a rural township thirty-seven kilometres north-west of Bendigo. It was formerly called Bridgewater-on-Loddon, being the location of one of the earliest bridges across the river.

Bridport *T* was first shown on George Frankland's map dated 1837. It no doubt derived from the place in the County of Dorset in England, as was the practice with many other Tasmanian place names.

Bright *V*, an alpine resort township seventy-six kilometres south-east of Wangaratta, was named after John Bright, the English orator and statesman of the nineteenth century. Originally named Morses Creek after prospector W. Morse, the township was renamed Bright in August 1862.

Brighton *SA* is the name of two places in South Australia.

1. Brighton is a suburb and corporate city seventeen kilometres south-west of Adelaide. Like its English namesake, it is a seaside resort. Brighton was first examined by Colonel Light in about 1837. The first sections were surveyed in 1838 and allotted in 1839. In the following year, several sections were surveyed into allotments by the surveyor, J.H. Hughes, for Matthew Smith, a solicitor and land agent, who named it Brighton after the English Brighton. Newspapers of the early 1840s referred to the name, Brighton. The earliest recorded plan of the subdivision referring to the name Brighton is dated 24 June 1848. The surveyor was R.I. Loveday.

In the 1840s, two of Brighton's 'industries' were whaling and smuggling. The smuggling chiefly comprised tobacco and spirits, which were landed by American whaling ships at Kangaroo Island and transferred, at the first suitable opportunity, to some of the bays along the coast from Glenelg to the mouth of the Onkaparinga River. 'Brighton House', North Brighton, is now part of Minda Homes Incorporated. It was built in the early 1840s by Michael Featherstone, an early colonist, who arrived in Adelaide with his family in 1839. Featherstone donated land for the first church built in Brighton, and the cemetery was established there in 1843.

2. Brighton is also the name of a private subdivision laid out in 1875 just north of Port Hughes about 160 kilometres north of Adelaide.

Brighton *T* was named by Governor Lachlan Macquarie in 1821. Although it has puzzled many that he should have applied the English seaside town's name so far inland, it is probably that he likened the area to another Brighton. There is a small village named Brighton in Fifeshire, Scotland.

Brighton *V* is a city on Port Phillip Bay in the Melbourne metropolitan area. For a short time it was called Waterville. Then it became Brighton after the town in England. Brighton's first settler, Henry Dendy arrived in 1841 and took up a grant of land. Dendy intended to set up an English-type estate with himself as squire. However, his affairs did not go well and he was eventually forced to sell his whole estate. The borough of Brighton was created in 1859. Brighton became a city in 1919.

Brighton-le-Sands *NSW*, a suburb of the Municipality of Rockdale, was originally called New Brighton and was named by property developer Thomas Saywell who modelled it on the English Brighton.

Bringelly *NSW* is an Aboriginal word meaning 'unobtainable'.

Brisbane *Q* was named in honour of Sir Thomas Macdougall Brisbane, who succeeded Macquarie as Governor of New South Wales in 1821. It is the only state capital to be named after a governor.

The Brisbane River was discovered, or rediscovered, during Brisbane's term of office. It was discovered on 2 December 1823 by John Oxley who had been ordered to find a suitable locality for the establishment of another penal colony. During October and November, Oxley explored Port Curtis and inspected Moreton Bay, where he met a white man who was living there with the Aboriginals. He went up the river for about seventy kilometres, was greatly impressed by it and named it after the Governor.

In company with Lieutenant Miller, a detachment of soldiers and a number of convicts, Oxley landed at Redcliffe in September 1824, but recommended the present site of Brisbane as a more suitable locality, because it offered good, fresh water and a safe anchorage. In December of that year the site was inspected by the Chief Justice, Sir Francis Forbes, and named by him Edinglassie. The penal settlement was finally established at Edinglassie in 1825, but eventually the name, Brisbane, prevailed.

The move upstream was made about May 1825. The convicts landed on a spit that the Aboriginals called

Meantjin, or Meganchin, somewhere near where Brisbane's Customs House was later built.

Brisbane Street Names The main streets of Brisbane are named after kings and queens of England. Roma Street is named after the wife of Governor Bowen, who was Countess of Roma. Wickham Terrace is called after Captain John Clements Wickham, who arrived in Brisbane in 1843 to head the administration as police magistrate.

Brisbane Water *NSW* was named after Governor Brisbane and the first surveys for land were made in 1825. Settlement began around Brisbane Water at this time.

Broadford *V* is a township by an old ford on Sunday Creek, seventy-four kilometres north of Melbourne. It was named after Broadford in Scotland.

Broadmeadow *NSW* is a locality in the Newcastle area. The name is descriptive of the area.

Broadmeadows *V* is a city with a military camp about twenty kilometres north of the centre of Melbourne. Broadmeadows is named after an early hotel in the district which was in turn derived from an estate in Berwickshire. Broadmeadows was created a district in 1857. It became a city in 1956.

Broadview *SA* is a suburb in the City of Enfield in the Adelaide metropolitan area. Broadview is a descriptive name bestowed because it occupies a high position in the northern suburbs. It commanded a fine view of the Adelaide Plains when it was laid out by C.H. Angas and K.D. Bowman in1918. Later housing development has rather obstructed the view. Broadview is adjacent to the suburb of Clearview.

Brodribb River *V* is a rural area eleven kilometres east of Orbost by Brodribb River. It was named after William Adams Brodribb, an explorer and pastoralist who established a station there in 1841.

Broke *NSW* is called after Sir Charles Broke Vere. It was named by the explorer T.L. Mitchell, who was acquainted with him.

Broken Bay *NSW* was the first European name connected with the Barrenjoey Peninsula. James Cook sailed from Botany Bay on 6 May 1770 and at sunset, he named broken land 'that seemed to form a bay'. The land he passed must have been the beach between present day Collaroy and Narrabeen. The Broken Bay of today was discovered by Governor Phillip on his visit of 2 March1788.

Broken Hill *NSW* In September 1883 a boundary rider pegged out sixteen hectares of the 'broken hill' which he believed to be a source of tin but which proved to be an outcrop of ironstone. The Broken Hill Proprietary Company was floated two years later. The Aboriginal name was Willyama, 'hill with a broken contour', and even in 1886 a Mines Department report said that this was the official name of the township. Eventually the European name survived, which seems a pity, especially as it has the same meaning as the Aboriginal name.

Brompton Park *SA* was a suburb of Hindmarsh in the inner metropolitan area of Adelaide. Brompton Park was named by Patrick Boyce Coglin, a mayor of Hindmarsh, who subdivided the area in 1875. He laid out Croydon as well as Brompton Park. Both are names of areas in and around London. Coglin's name is perpetuated in the Town of Coglin and four streets in the metropolitan area. Brompton Park is now included in the suburb of Brompton which was subdivided and sold in 1849 by Nathaniel Hailes.

Bronte *NSW*, located in the Municipality of Waverley, is one of Sydney's Eastern Suburbs. Bronte gets its name from 'Bronte House', which was built at Nelson Bay by Robert Lowe (1811-1892), a British barrister who arrived in Sydney in 1842. Lowe bought the land from Mortimer Lewis, who had already laid the foundations for a house but was forced to sell because of the economic depression at the time. Lowe had the house built in1845 or 1846. Lowe returned to England in 1850 after a turbulent career in Australia. He became a member of the House of Commons, a Chancellor of the Exchequer, and finally ended his days as Lord Sherbrooke. Lowe did not bestow the name Bronte on the house. The house still stands. It was purchased by the local council in1948 to ensure its preservation. In 1983, the council leased it to a wealthy businessman in return for a guarantee to restore it to its former glory.

Brooklyn *NSW*, a locality within Hornsby Shire on the North Shore of Sydney, got its name because the Union Bridge Company of the United States of America won the tender to build the first rail bridge over the Hawkesbury. The company, which also built the New York Brooklyn Bridge completed the Hawkesbury bridge in 1889. The railway brought tourists to the area. In 1901, the Duke and Duchess of York toured the river by ship.

Brookvale *NSW* is a suburb of Warringah Shire. Brookvale was originally called Greendale by William Frederick Parker, who bought land in the district in 1836. His son, William Francis Parker, built 'Brooklands' in 1879. The old home can still be seen behind Warringah Mall. The name was changed from Greendale to Brookvale by the postal authorities at a much later date, to avoid confusion with a little hamlet of the same name in the Bringelly district.

Broome *WA* was named after Sir Frederick Napier Broome, a governor of Western Australia. Pearlers established a settlement on the site in 1883. The town-site was declared on 27 November 1883.

Broughton, Port *see* **Port Broughton**

Brown Hill Creek *SA* takes its name from the colour of the hill which is its source. The name occurs early in the history of white settlement in the Adelaide area. It is shown as Brown Hill Rivulet on Light's map of 1839. The Aboriginal names for the creek are Willawilla or Wirraparinga.

Bruce *ACT* is named in honour of Lord Stanley Melbourne Bruce (1883-1967) who was Prime Minister of Australia from 1923 to 1929.

Bruce, Mount *see* **Mount Bruce**

Brunswick *V* is an industrial city to the north of Melbourne. Brunswick got its name because Thomas Wilkinson, who bought land there in 1841, called his estate after Princess Caroline of Brunswick, the wife of King George IV. She had been denied her rights as Queen and Wilkinson had been involved in the campaign waged on her behalf. Brunswick was proclaimed a municipal district in 1857. In 1888, it was proclaimed a town. It became a city in 1908.

Bruny Island *T* was named by the French Admiral, Bruny (or Bruni) D'Entrecasteaux in 1792. D'Entrecasteaux's map shows it as Ile de Bruny.

Bruthen *V* is a rural township twenty-four kilometres northeast of Bairnsdale by the Tambo River. It is an Aboriginal word meaning 'place of evil spirit'. It was formerly called Tambo.

Buchan *V* is a rural township north-west of Orbost by the Buchan River. It is an Aboriginal word meaning 'grass bag'.

Bucketts *NSW* is a name taken from the Aboriginal word meaning 'big rocks'.

Buckland Hill *WA* is a feature of Mosman Park in the Perth Metropolitan area. The name was first recorded by George Smythe during surveys in January 1833 as 'Buckland Downs, Ditto Hill'. The locality around the hill was shown as Buckland Downs on a map of the colony drawn in London for the *Journal of the Royal Geographical Society* by John

Murray in September 1832. The name, Buckland Downs, is thought to have been bestowed in 1827 by Governor Stirling to honour William Buckland, the first reader in geology at Oxford University and later Dean of Westminster, but this has not been established with any certainty. The name, Buckland Downs, was gradually superseded by Buckland Hill which came to be applied to both hill and locality for a time. The locality is now the Town of Mosman Park but the hill retains its original name.

Buderim *Q* is an Aboriginal word for 'honeysuckle'.

Budgewoi *NSW* means 'soft bark tea tree'.

Buffalo Creek *NSW*, a stream that flows through part of the Sydney suburb of Ryde, commemorates HMS *Buffalo*, on which Captain Raven, the owner of land at Ravens Points, had sailed.

Buffalo, Mount *see* **Mount Buffalo**

Bugle Range *SA* has two possible derivations. One is that a bullock called 'Bugle' used to wander from home into the ranges, and the second is that an early German settler used to make the gullies ring during the summer evenings with the notes of his bugle.

Bulahdelah *NSW* means 'the meeting of the waters'. It is pronounced Bull-er-dealer (rhymes with pull).

Bulla *V* is a mainly agricultural shire on the outskirts of the Melbourne area. Bulla means 'two' in the local Aboriginal language, so *bulla bulla* could refer to four landmarks—four hills or four creeks. Bulla was created a district in 1862. It was proclaimed a shire in 1866.

Bullaburra *NSW*, a locality in the Blue Mountains west of Sydney, was named by Arthur Rickard who subdivided the locality. The railway station was opened on 16 February 1925.

Buller, Mount *see* **Mount Buller**

Bulli *NSW* The original Aboriginal name was Bulla or Bulla Bulla, meaning 'two mountains' (Mounts Kembla and Keira). As early as 1823 reference was made to a small landholding at Bull Eye. Other more fanciful meanings for Bulli have been given, e.g. 'white grubs', and 'place where the mooki (Christmas bush) grows'.

Bullumwaal *V* is an old goldfields settlement twenty-eight kilometres north of Bairnsdale. It is an Aboriginal word meaning 'two spears'.

Bumbaldry *NSW* is an onomatopoeic Aboriginal word to imitate the sound of people jumping into water.

Bunbury *WA* was inspected in 1836 by Governor Stirling and Lieutenant H.W. Bunbury. Stirling decided it was a suitable town-site and named it in honour of Lieutenant Bunbury. The town was proclaimed in 1840 and surveyed in 1841.

Bundaberg *Q* was chosen as a town-site by District Surveyor Thompson in 1869. Thompson also chose the name Bundaberg. 'Bunda' came from the Aboriginal tribe who had adopted Thompson's assistant, Edwards, into their tribe. 'Berg' means 'town', so Bundaberg—Bunda's town—was born. Thompson also chose the name, Bourbon, for the street that became the town's main thoroughfare. Its origin is uncertain. It may come from an Aboriginal word *boorbung* meaning a 'chain of waterholes', or Bourbon, a variety of sugar cane. Bundaberg became a town in 1867 and a city in 1913.

Bundalong *V* is a rural township east of Yarrawonga by Lake Mulwala near the junction of the Ovens and Murray Rivers. It is an Aboriginal word meaning 'joined'.

Bundanoon *NSW* comes from an Aboriginal word meaning 'place of deep gullies'.

Bundeena *NSW*, in Sutherland Shire, is an Aboriginal word meaning 'noise like thunder'.

Bundoora *V* is a locality in Diamond Valley and Whittlesea Shires on the outskirts of the Melbourne metropolitan area. Bundoora was so named because the missionary George Langhorne told Surveyor Robert Hoddle that Keelbundoora was the name of a local tribesman and the eldest son of Jika Jika, one of the Aboriginal chiefs that made the treaty which was supposed to sell the site of Melbourne to John Batman in 1835. Bundoora is said to mean 'many kangaroos'.

Bungan Beach *NSW* in Warringah shire was named after Bungan Head but no information is available on the derivation of the name. It was possibly confused early with Bongan Beach, an old name for Mona Vale Beach.

Bungaree *V* is a rural township ten kilometres east of Ballarat. It is the name of one of the Aboriginals who signed the document conveying land at Melbourne to John Batman.

Buninyong *V* is a rural township eleven kilometres south of Ballarat at the base of Mount Buningyong.

Bunnerong *NSW*, a locality in the Municipality of Randwick, is an Aboriginal name meaning 'small creek'. 'Bunnerong

House' was the first dwelling in the municipality. It was built at Bunnerong in 1825 by John Neathway Brown, a military veteran who received the first of several grants given in the area. It was a stone residence of eleven rooms surrounded by orchards and an orange grove. It was later the site of Bunnerong Power Station.

Bunya Mountains *Q* An Aboriginal name for a species of pine that is plentiful in the mountains. They were an important gathering place for Aboriginal tribes from a great distance, as the edible nuts of the pine were valued as a food supply.

Bunyan *NSW* comes from an Aboriginal word meaning 'place where pigeons stay'.

Bunyip *V* The Aboriginal people's legendary monster of lakes and swamps, of which grim and unusual tales are told. The hollow booming so often heard from the margin of reedy swamps, more hollow and louder by night than by day, was the voice of the bunyip—even though sceptical white men might ascribe it to the lonely bittern. The Aboriginals' belief in such powerful spirit beings as the Rainbow Serpent, which dwelt in waterholes and rivers, may be the basis of legends of the bunyip.

Burdekin River *Q* was originally named Wickham River by Captain Wickham during his exploration of the northern coastline in the early 1840s. Leichhardt visited the area in 1845 and renamed the river in honour of Ann Burdekin, the wife of Thomas Burdekin of Sydney who had assisted him with his expedition.

Burketown *Q* After Robert O'Hara Burke, leader of the ill-fated Burke and Wills expedition.

Burleigh Heads *Q* was named by the surveyor J.R. Warner in 1840. It was originally spelt as Burly Heads. Little Burleigh had the Aboriginal name Jellurgul and Big Burleigh, Jabbribillum.

Burnett River *Q* Named after James Charles Burnett, the surveyor who discovered it in 1847. Burnett was a clerk under Sir Thomas Mitchell. He rose to be surveyor in charge of the department in Queensland and explored the Mary and Burnett rivers. The Aboriginal name for the upper reaches was Borrallborrall, and of the lower, Birrabirra.

Burnie *T* was named after William Burnie, a director of the Van Diemen's Land Company, which opened land to tenants in 1842 and marked out a town reserve at the western end of Emu Bay.

Burning Mountain *NSW*. A common name for Mount Wingen, the Aboriginal name which means 'fire'. Mount Wingen is one of the natural curiosities of Australia. There is a seam of coal within the mountain which has been burning slowly, probably for thousands of years, causing smoke to rise through fissures in the ground. There are also sulphur deposits which cover the hot rocks with a yellow coating. Since the phenomenon was observed in 1828 (and again by T.L. Mitchell in 1831) the burning area has crept less than a hundred metres up the slope. The slow rate of combustion is due to lack of oxygen deep under the surface. According to Aboriginal legend, it was caused by a tribesman who was lighting a fire on the mountain one day when he was carried off deep into the earth by 'the evil one'. Unable to escape, he used his firesticks to set the mountain alight, so that the smoke might warn others to keep away.

Burnley *V* is a suburb of Richmond, one of Melbourne's oldest inner suburbs. Burnley is called after the first landowner, a lawyer named William Burnley.

Burns Bay *NSW*, a locality in the Municipality of Lane Cove, was originally called Murdering Bay. It owes its present name to one of three people associated with the history in this area. Barnet Burns had an orchard in the district but it was at the head of Tambourine Bay and in any case he sailed off to New Zealand in 1824 and never worked his grant at Lane Cove. Another contender was Captain George Bunn, who tried unsuccessfully to set up a whaling station at Lane Cove. However, the Burns of Burns Bay was most likely Terrence Byrne, or Burn, an Irish convict who squatted on land near the boundary of the present Burn Bay Park. In 1827 he applied for the land to be granted to him. While waiting for the authorities to process his application he cleared and cultivated much of the area. Five years later the authorities refused his application for a grant, but told Byrne he could purchase ten acres (4 hectares) at five shillings an acre. He refused and walked off the land he had already cleared.

Burnside *SA* is a corporate city on the south-eastern outskirts of the Adelaide metropolitan area. The local history of Burnside says the name came from the farm of the Anderson family of East Lothian, Scotland, who arrived in Adelaide in 1839 and took up land in the district. Peter Anderson called his farm Burnside because of the burn (creek) which flowed beside his property. The South Australian Company laid out the township in 1849 as 'Burnside the Beautiful'.

Burra *SA* was the site of a copper mine which was discovered in October 1845 on the Burra Creek. The creek was named by Indian workers in the employ of Mr James Stein who held country there. It is Hindustani for 'great great'. The Aboriginal name was Kooringa. The rich copper discovery was made by Thomas Pickett, a shepherd employed by Stein. More copper was later found at the northern end of F.H. Dutton's property. In the heyday of the mine, the adjoining settlement was South Australia's largest country town, but it was not called Burra until the railway set up its station sign in 1870. The largest part of the town, the business section south of the mine, was Kooringa (Aboriginal for 'she-oak' and 'creek'), a name that still persists in some quarters. North of the mine, the settlements were known as Aberdeen and Redruth and were separated by the Burra Creek. In addition, there were suburbs such as Copperhouse and Hampton. The government town of Redruth was surveyed in 1849 and named after a small town and parish in Cornwall. It was altered to Burra in September 1940. The private towns of Kooringa, Llwchwr, Aberdeen, New Aberdeen, and Graham were also altered to Burra in the *Government Gazette* dated 19 September 1940.

Burraneer Bay *NSW* in Sutherland Shire, was probably named by Lord Audley when he made the first official survey of Port Hacking and the Hacking River in 1863-1864.

Burrinjuck *NSW* An Aboriginal term for 'precipitous' or 'rugged-topped mountain'.

Burrowye *V* is a rural area north-west of Corryong. It is an Aboriginal word meaning 'eastwards'.

Burrum *Q* is the name of a river and a shire between Maryborough and Bundaberg. The name comes from an Aboriginal word meaning 'big'.

Burrumbeet *V* is a rural area twenty-two kilometres north-west of Ballarat by Lake Burrumbeet. It is an Aboriginal word meaning 'muddy water'.

Burwood *V* is a suburb of Camberwell, a city in the Melbourne metropolitan area. Burwood takes its name from a mansion built in the early 1850s by Dr James Palmer, a physician, cordial maker, squatter, wine merchant and early mayor of Melbourne.

Bushrangers Hill *NSW*, a locality in Warringah Shire, got its name from a trig station called Bushranger. The trig station may have been so named because you can see up to Palm Beach in one direction and to Narrabeen in the other—very useful as a vantage point for bushrangers on the run.

Busselton *WA* was settled by John Garrett Bussell and his three brothers in 1832. They named their run 'Cattle Chosen'. The district was known by several names—The Vasse, Cattle Chosen and Bussell Town. But when the site was declared a town in June 1837 the name became stabilised as Busselton. In 1876, Grace Bussell became a local legend when she and an Aboriginal stockman named Sam Isaacs plunged into the breakers on horseback to rescue fifty people from the shipwrecked *Georgette*.

Bustard Bay *Q* was named by Captain James Cook on 23 May 1770 because of the bustards he found there. Cook's men shot one of the birds, which was about the size of a turkey. Cook noted in his journal: 'We all agreed that this was the best bird we had eaten since we left England and in honour of it we called this inlet Bustard Bay.'

Buxton *V* is a rural locality forty kilometres north-east of Healesville at the junction of the Acheron and Steavenson Rivers. It was named by Surveyor Darbyshire after Buxton Springs in Derbyshire, England.

Byaduk *V* is a farming area twenty-six kilometres south of Hamilton. It comes from an Aboriginal word meaning 'stone axe'.

Byford *WA* was originally called Beenup which was a corruption of the name of Beenyup Brook. In 1906, Surveyor A.W. Canning commenced laying out a number of town-site lots in the area and 'Beenup Townsite' was proclaimed in May the same year. By 1919 dissatisfaction with the town's name was general amongst the residents and the Progress Association requested that the town be renamed Lynwood. This request was refused because of a duplication in Queensland. However the residents, determined to have the town renamed, submitted the name Byford. The reasons for choosing this name are now unknown but the proposal was agreed to by the authorities and the town was officially renamed Byford on 23 April 1920.

Byron Bay and **Cape Byron** *NSW* were named by Captain Cook after Captain John Byron, grandfather of the poet Byron. Captain Byron circumnavigated the globe in 1764-66 and thus preceded Cook in the Pacific. He was in command of HMS *Dolphin*. On 15 May 1770, Cook wrote: 'A Tolerable high point of land bore N.W. by W., distant 3 miles; this point I named Cape Byron.' The bay was also named at this time. Cape Byron is the most easterly point of the continent.

C

Cabarita *NSW* is a locality in the Sydney municipality of Condord. It is an Aboriginal word meaning 'by the water'. Its origin as a place name is uncertain. It may have been bestowed by Sir Thomas Mitchell who, as Surveyor-General, had the stated policy of retaining Aboriginal names wherever possible. It could also have been suggested by either the poet Henry Kendall, who was a clerk in the Surveyor-General's office in Sydney from 1863 to 1866, or by his uncle, Captain Joseph Kendall who was a whaling master. A bay on the western side of Cabarita is named Kendall Bay.

Caboolture *Q* comes from a Kabi tribe word, meaning 'place of carpet snakes'.

Cabramatta *NSW* is a suburb of Fairfield, a city on the outskirts of the Sydney metropolitan area. The name is derived from an Aboriginal word. Some authorities translate it as *cabra* meaning 'white grub' and *matta* meaning 'a point of land'. Other possible meanings include 'place of good fishing' or 'head of waters'. The name was first applied by Mr N.G. Bull, a prominent politician, who called his estate 'Cabramatta Park'. It is interesting to note that as early as 1795, the Cabramatta district was referred to on maps as the Moonshine Run. This title is said to have been originated by an early settler who expressed the view that timber on the run was so thick that moonshine could not penetrate the foliage.

Cadibarrawirracanna, Lake *see* **Lake Cadibarrawirracanna**

Cairns *Q* lies on Trinity Bay over 1600 kilometres north of Brisbane. Trinity Bay was declared a port of entry and clearance on 1 November 1876 and named Cairns after the then Governor of Queensland, Sir William Cairns. On that day, Captain Lake of the SS *Victoria* invited all hands to lunch on his ship and the town was duly christened. The locals pronounce this as 'cans' as in tin cans, not as in 'cairns', a memorial pile of stones.

Calga *NSW* means 'a stony ridge'.

Calista *WA*, a residential suburb of Kwinana, a town near Perth, takes its name from the ship *Calista*, which arrived in Western Australia with seventy-three passengers on 5 August 1829. The suburb was named in 1953.

Callabonna, Lake *see* **Lake Callabonna**

Calliope River *Q* is named after HMS *Calliope*, which carried Governor Sir Charles Fitzroy to Gladstone to swear in Sir Maurice Charles O'Connell as Government Resident in 1854.

Caloundra *Q* is a tourist resort on the Sunshine Coast ninety-six kilometres north of Brisbane. The Ngundanbi Aboriginals are thought to have occupied the area before European settlement. The name is thought to derive from the Aboriginal word *callanda*, meaning 'beautiful'.

Calwell *ACT* is called after Arthur Calwell (1896-1973), leader of the Labour Opposition from 1960 to 1967.

Camberwell *V* is a residential city about nine kilometres east of the centre of Melbourne. The Camberwell area was originally an Aboriginal hunting ground. Surveyor Robert Hoddle named it Boroondara, an Aboriginal name meaning 'shady place'. The first white settler was a pastoralist named John Gardiner, who took up land in the district in 1837. Camberwell got its name because in 1853 George Eastaway built a hotel of corrugated iron on the north-east corner of what is now Camberwell Junction. In searching for a suitable name, he thought of the roads that converged at Camberwell Green in his native London. He called the hotel Camberwell Inn. Local government began with the Boroondara Roads Board, which was proclaimed in 1854. In 1871, Boroondara Shire was proclaimed. In 1902, the name was changed to Camberwell and Boroondara Shires. Camberwell was proclaimed a town in 1906, and became a city in 1914.

Camden *NSW* was officially named after Lord Camden, Secretary of State for the Colonies in 1820. Its Aboriginal name is Benhennie, meaning 'the dry land'.

Camellia *NSW* is located on the southern side of the Parramatta River and was named after Silas Sheather's Camellia Grove Nursery in the area.

Cammeray *NSW*, a locality in North Sydney, is one of the few reminders of the original inhabitants of the area. The northern shore of Port Jackson from Lane Cove River to Middle Harbour belonged to the Camaraigal, feared warriors who gave their name to the suburb.

Camooweal *Q* is a small pastoral centre on the Georgina River. The name, Camooweal, is a compound of 'camel' and 'Weale'. The first camels in the district are said to have been introduced by a surveyor named Weale when the town was surveyed. The town was first gazetted in 1884.

Camp Cove *NSW* is located on Sydney Harbour in the

Municipality of Woollahra. Camp Cove was the site of Governor Phillip's first landing within Port Jackson when he explored Sydney Harbour from Botany Bay in 1788. An obelisk and tablet now commemorate the landing. The name Camp Cove has been used from the earliest days of settlement when harbour pilots camped there taking advantage of the sandy beach and the shelter for small boats.

Campaspe River *V* Discovered and named by T.L. Mitchell in October 1836. Campaspe was a favourite concubine of Alexander the Great.

Campbell *ACT* is named after Robert Campbell (1769-1846), a pioneer and merchant, who took up a grant in 1825 and built Duntroon homestead, which is now part of the Royal Military College at Duntroon.

Campbell, Port *see* **Port Campbell**

Campbell Town *T* was named by Governor Macquarie, during his visit in 1821, after his wife's maiden name.

Campbellfield *V* is a suburb of Broadmeadows, a city in the Melbourne metropolitan area. Campbellfield is named either after Neil Campbell, a well-known landowner and grazier in the district, or Charles and James Campbell who had a station in the district in the 1840s.

Campbelltown *NSW* is a satellite city fifty kilometres south of Sydney. It was named by Governor Macquarie when he revisited the district in 1820 to mark out a town-site. Campbell was his wife's maiden name.

Campbelltown *SA* is a corporate city in the Adelaide metropolitan area. Campbelltown is named after Charles Campbell, who bought a large section of land in the district in 1842.

Camperdown *NSW* is a locality in the City of Sydney. Camperdown recalls the naval battle of that name fought in 1797. The battle was fought off the Dutch coast near the village of Camperdown. William Bligh named his estate after the battle in which he commanded the sixty-four-gun ship *Director*.

Camperdown *V*, a town at the foot of Mount Leura, forty-five kilometres west of Colac, was named in 1854 on La Trobe's suggestion because the oldest settler when the township was laid out was Duncan, which was also the name of the British commander who was victorious in a great naval battle with the Dutch in 1797 and was made Earl of Camperdown.

Campsie *NSW* is a suburb of the Sydney municipality of

Canterbury. Campsie was subdivided in 1885 as the Campsie Estate and the streets were dedicated by the owners, the Anglo-Australian Company, in December of that year. Development was slow at first and Campsie did not really become a fully developed shopping and residential area until about 1925. It is named after a range of hills in Stirlingshire, Scotland.

Canada Bay *NSW*, a locality in the Sydney municipality of Concord, together with adjacent Exile and France Bays, recalls an unusual event in Commonwealth history. In 1837 there were rebellions against British rule in Canada. In 1840, fifty-eight French Canadians were exiled to New South Wales. They worked at quarrying in the bays that now recall their presence. Of the original fifty-eight, two died, and one married an Australian girl and settled in Dapto. The others were pardoned and eventually most of them returned to Canada.

Canberra *ACT* is derived from an Aboriginal word, and is usually translated as 'meeting place', which would be an appropriate name for the site of a national parliament, if this was definitely the derivation. Unfortunately, this is not so. Like most Aboriginal names, the meaning is the subject of conjecture and dispute. Another suggested meaning is 'a woman's breasts', which would describe the hilly landscape around Canberra. Aldo Massola, a recognized authority on Aboriginal languages, says that Canberra comes from *nganbirra* meaning a camping place. He says that the first Aboriginal name recorded was Pialligo, a corruption of *paialla'bo*, which means 'I'll let you go by'. This was the answer given by an Aboriginal to Ainslie, a drover for the first squatter, when asked for the name. Ainslie put this name on record. However, Europeans generally referred to the area as Limestone Plains during the early years of settlement. In 1824, convicts and an overseer employed by Joshua Moore built huts and established a stock station where the suburb of Acton now stands. It was located near the site of the present Canberra Hospital. Moore named his stock station 'Canbury', which he believed was the Aboriginal name of the place. Other early European references give the name as Canberry, Kembery and Gnabra. Ninety years later this name, which had evolved with usage into Canberra, was chosen for the national capital.

Caniambo *V* is a rural area south-east of Shepparton. It comes from Aboriginal words meaning 'place for blackfellow'. It was formerly known as Gowangandle.

Canley Vale *NSW*, is a suburb of Fairfield on the south-western outskirts of the Sydney metropolitan area. It was named by Sir Henry Parkes, who built a mansion-style house adjacent to the site of the present railway station. Parkes called it after his birthplace Canley Grange, in Warwickshire, England. Canley Heights derived from Canley Vale.

Cann River *V* is the name of a river and a township. The river, which rises in the Coast Range and flows through Lake Furnell into Tamboon Inlet, was named after Cann, who was a squatter in the district in the 1850s. The rural township was named after the river.

Canning *WA* is a city in the Perth Metropolitan area. It takes its name from the Canning River which flows through the city. The river itself was named in 1827 in honour of George Canning, then British prime minister.

Canning Stock Route *WA* The route, which gave an outlet for cattle from the Kimberleys to the railway at Wiluna, was explored by the surveyor, Alfred Wernham Canning in 1906, and also pioneered by H.S. Trotman from 1907 to 1910. Canning was also notable for his survey for Western Australia's first rabbit-proof fence.

The 1400 kilometre-long stock route is negotiable only in good years. It was used during World War II, but has since fallen into disuse. The route was originally planned to enable East Kimberley pastoralists to take tick-infested stock south without passing through the tick-free West Kimberley. It was also believed that ticks would be unable to survive conditions in the dry and arid country through which the stock route passed. Originally named the Wiluna Kimberley Stock Route it was commonly called the Canning Stock Route. The name was formally approved by the Surveyor-General on 3 February 1967.

Canobolas, Mount *see* **Mount Canobolas**

Canowindra *NSW* is Aboriginal for 'home' or 'camping place'.

Canterbury *NSW* is a municipality in the Sydney metropolitan area. It was named by the first land grantee in the district, the Reverend Richard Johnson, who was also the first clergyman to serve in New South Wales. The grant was made in 1793 and Johnson named it Canterbury Vale.

Canterbury *V* is a suburb of Camberwell, a city in the Melbourne metropolitan area. It was named after John Henry Manners-Sutton, Viscount Canterbury, who was Governor of Victoria from 1866 to 1873. The name Canterbury Heights Estate was selected by land auctioneers when the orchard area was sub-divided for housing in 1888. Streets were named after English counties.

Canungra *Q* is an Aboriginal word for 'small owl'.

Capalaba *Q* comes from an Aboriginal word meaning 'place of possum scrub'.

Cape Adieu *SA* is the point at which Nicolas Baudin completed his charting of the coast. Baudin named this cape in his log as Pointe du Depart, 'Departure Point'. Following his death on the return voyage, subsequent publication of journals and charts by Freycinet and Peron referred to it as C. des Adieux. It is now spelt as Cape Adieu.

Cape Barren Island *T* takes its name from Cape Barren, which is located at the island's eastern extremity. The cape was sighted and so named on 18 March 1773, by Tobias Furneaux in the *Adventure*. The island and the main settlement on it subsequently took their names from the cape, although the settlement is simply known to the locals as The Corner.

Cape Byron *see* **Byron Bay**

Cape Catastrophe *SA* The nomenclature of this and other natural features in the vicinity of the entrance to Spencer Gulf was chosen by Matthew Flinders to commemorate the tragedy when eight of his crew were drowned on the evening of 21 February 1802. The *Investigator* passed Cape Wiles. Flinders sent a cutter ahead to seek a safe anchorage for the vessel. The sailing master, or mate, John Thistle, was in command, together with William Taylor, midshipman, and a crew of six. On its return the cutter was sighted and then disappeared in a sudden squall. A search proved in vain. The following day broken pieces of the boat and a cask which had been filled with fresh water were found, but there was no trace of the bodies.

'I proceeded to the southern extremity of the main land, which was now named Cape Catastrophe' wrote Flinders. He called the inlet where the party had been ashore for water, Memory Cove, and named islands in the vicinity after the men who were drowned.

Cape Grim *T* On 9 December 1798, Flinders and Bass in the *Norfolk* were sailing along the northern coast of Van Diemen's Land. Their intention was to confirm the theory that it was an island. When they noticed that the coast was turning southwards and that a long swell was rolling in from the west, they realised that they had come through a strait, thus proving that Van Diemen's Land was an island. The appearance of the headland caused them to name it Cape Grim. The Aboriginal name was Kennaook.

Cape Howe *V* was named by Captain Cook on 20 April 1770.

This south-east limit of the mainland was named after Admiral Richard Howe, the first Earl Howe, who was treasurer of the Royal Navy at that time.

Cape Keer-Weer *Q* Cabo Keerweer, 'Cape Turnagain', was the first European name bestowed on any part of Australia. The *Duyfken*, under the command of Captain Willem Jansz, and with subcargo Jan Lodewijs van Roosengin, sailed south-east from the coast of New Guinea and in 1606 coasted down the western shore of Cape York Peninsula. Running short of water and provisions, the captain turned back after reaching 'Cape Turnagain'. While surveying the Gulf of Carpentaria in 1802, Flinders confirmed the Dutch name 'from respect to antiquity'.

Cape Leeuwin *WA* was named in 1801 by Matthew Flinders after the Dutch ship *Leeuwin*, 'Lion', that had sailed along the coast in 1622.

Cape Naturaliste *WA* was called after the *Naturaliste*, a support vessel of Nicholas Baudin's French Scientific Expedition, which landed there in 1801. It was commanded by Captain Hamelin.

Cape Otway *V* is located at the north-western limits of Bass Strait. It was named by Lieutenant James Grant in honour of his friend, Captain William Albany Otway RN.

Cape Schanck *V*, a headland south-west of Point Nepean on the Mornington Peninsula, was named by Lieutenant James Grant when on the *Lady Nelson* in 1802, in honour of Admiral John Schanck. He had designed the *Lady Nelson* and was the inventor of the sliding keel with which it was equipped.

Cape Solander *NSW* in the Sutherland Shire was named Point Solander by Captain Cook for Dr Carl Solander, a Swedish botanist aboard the *Endeavour*.

Cape Tribulation *Q* was named on 10 June 1770, the day before the *Endeavour* grounded on a reef. Cook named the northern point he had in sight 'Cape Tribulation, because here began all our Troubles'.

Cape Woolamai *V* is the south-eastern tip of Phillip Island. It was named by George Bass in 1798. It is an Aboriginal word meaning 'snapper'.

Cape York Peninsula *Q* was named by Captain Cook after the Duke of York in 1770.

Capital Hill *ACT* got its name because Canberra is the federal capital of the Commonwealth of Australia.

Captains Flat *NSW* is said to have been named after a bullock which grazed on the banks of the Molongolo River.

Carboor *V* is a rural area south-west of Myrtleford. It is an Aboriginal word meaning 'koala'.

Carbrook *Q* was called Tablabuba or Tabooba meaning 'bitter water' by the Yaggapals, the local Aboriginals. German immigrants, who arrived in the 1860s, called it Gramzow after a town in Germany. The name was changed to Carbrook in 1916, because of anti-German feeling during World War I.

Carcoar *NSW* An Aboriginal word probably meaning 'kookaburra'. The syllables imitate the laughing cry—cah-co-ah. It is just possible that it may instead (or in addition) be a rendering of the voice of the frog or of the crow, both speculations having been made in the past.

Cardiff *NSW* is a locality in the Newcastle area. It is named after the city in Wales and means 'Fort on the River Taf'.

Cardinia *V* is a rural area north-west of Koo-wee-rup by Cardinia Creek. It is an Aboriginal word meaning 'sunrise'.

Careening Cove *NSW*, a locality in North Sydney, with its sheltered position and deep water frontages was used, as its name implies, for beaching and careening vessels.

Cargelligo, Lake *see* **Lake Cargelligo**

Caringbah *NSW* in Sutherland Shire is an Aboriginal word referring to the paddy-melon wallaby. Caringbah was originally called Highfield but the name was changed with the opening of the post office in 1912.

Carisbrook *V* is a rural locality seven kilometres east of Maryborough. It was named in memory of a daughter of an early settler. The name is really a condensed version of Caroline's Brook.

Carlingford *NSW* lies within the boundaries of the City of Parramatta. The Mobbs family had orchards there from before 1834. When a post office was opened in July 1883 it was named Mobbs Hill but on the representation of Mr A.H. McCulloch MLA the name was changed to Carlingford the next month. Carlingford is named after Lord Carlingford, Secretary of State for the Colonies from 1857 to 1860.

Carlsruhe *V* is a rural locality six kilometres south-east of Kyneton by the Campaspe River. It was named by Charles Elden after a town in Germany.

Carlton *NSW*, a suburb in the Municipalities of Rockdale,

Hurstville and Kogarah, was named after the original subdivision.

Carlton *V* is a suburb of Melbourne. It is named after the English residence of the Prince of Wales in London.

Carmel *WA* was named by Edward Owen who settled there in 1893. He established an orchard and founded a Methodist community there. This name was suggested by the South Kalamunda Progress Association in 1915 and the name was gazetted on 15 October 1915. *Carmel* is a Hebrew word and means 'vineyard or garden of God'.

Carnarvon *WA* was named after Lord Carnarvon, Secretary of State for the Colonies from 1866 to 1874. It was surveyed in 1882 and proclaimed in 1883.

Carnegie *V* is a suburb of the City of Caulfield in the Melbourne metropolitan area. Carnegie was the name chosen for the first railway station in 1904.

Carpentaria, Gulf of *see* **Gulf of Carpentaria**

Carrajung *V* is a rural area between Traralgon and Yarram. It is Aboriginal for 'fishing line'.

Carramar *NSW* is a suburb of Cabramatta on the southwestern outskirts of the Sydney metropolitan area. It is an Aboriginal expression meaning 'shade of trees'. The name was bestowed by the New South Wales Railway Department two years after the railway line from Regents Park to Cabramatta was opened in 1924. When first constructed, the railway station was called South Fairfield, but it caused so much confusion to train travellers (because Fairfield was on another railway route) that the name was changed to Carramar.

Carrington *NSW* is a coal shipment suburb near the entrance to the Hunter River. It is named after Baron Carrington, Governor of New South Wales (1885-1890).

Carrum *V* is a beach resort north of Frankston. Its name is possibly Aboriginal for 'yesterday'.

Carwarp *V* is a rural area seventeen kilometres south of Red Cliffs. It is Aboriginal for 'river bend'.

Cascades *T* is a very early name taken from the falls on the rivulet on which the original Degraves brewery was situated. The label of the renowned Cascade beer depicts one of the falls.

Casey, an Australian Antarctic Base in Wilkes Land, was set up in 1969. It was named after Lord Casey, who served as

Governor-General of Australia from 1965 to 1969.

Casino *NSW* is named after Monte Cassino, the Italian monastery-city. The station was established in the 1840s by Clarke Irving, and was named Cassino either by him or by Henry Clay and George Stapleton, two other early settlers. The modern spelling is a corruption.

Casterton *V* is a township sixty-three kilometres west of Hamilton by the Glenelg River. This area was taken up for settlement in 1837 and named, probably by one of the Henty brothers, from the town of this name in Westmoreland. It comes from an Old English name meaning 'camp'.

Castle Cove *NSW* is located on Sydney Harbour in the North Shore municipality of Willoughby near Castle Crag which gives the cove its name. The original land grant in this area was made to H.G. Alleyne in 1858. In the 1870s, H.C. Press acquired six hectares of land at Castle Cove and turned it into picnic grounds. On Sundays, ferries and boats would bring the merchant princes of Sydney and Mosman on picnics. Family parties of thirty or more were common, complete with their retinue of nannies, cooks, and butlers. When the era passed and people no longer picknicked in this style the Press family closed the grounds and chose Castle Cove as their home.

Castle Hill *NSW* is a locality in the Baulkham Hills Shire on the north-western outskirts of Sydney. It is of obscure origin. It is thought that Governor King, who was responsible for the establishment of the government farm there, chose the name to distinguish it from the farm at Toongabbie. When the known portions of the colony were first definitely divided into districts in 1821 the name Castle Hill was chosen for one of the areas.

Castlecrag *NSW* is a suburb of the Municipality of Willoughby on Sydney's North Shore. Castlecrag was originally named because of a large rocky outcrop on the peak of Edinburgh Road above the Tower Reserve. On early maps it was marked as Edinburgh Castle.

Castlemaine *V* is a city thirty-eight kilometres south of Bendigo. The area was crossed by Major T.L. Mitchell in 1836 and settled shortly afterwards. In 1851 gold was discovered at Specimen Valley. There was a sudden influx of miners to the diggings at both Forest Creek and Mount Alexander. It was named at the request of Governor La Trobe after Castlemaine in Ireland, because Viscount Castlemaine was the uncle of the chief goldfields commissioner, Captain W. Wright. Victorians pronounce this with a short 'a' as in 'ass', not a long 'a' as in 'farce'.

Castlereagh; Castlereagh River *NSW* On 6 December 1810 Governor Macquarie noted in his Journal: 'The township for the Evan or Nepean District I have named Castlereagh in honour of Lord Castlereagh.' The site first chosen by Macquarie is five kilometres from the present settlement. The Governor founded five towns in the district (Aboriginal name Mulgoa): Castlereagh, Windsor, Richmond, Wilberforce and Pitt Town. The Castlereagh River was discovered by G.W. Evans during Oxley's expedition in 1818 and was also named after Lord Castlereagh, Secretary of State for the Colonies, as was Castlereagh Street in Sydney.

Casula *NSW*, a suburb of Liverpool, occupies part of what were originally grants held by Dr Throsby (Glenfield), Richard Guise (Casula) and Captain Eber Bunker (Collingwood). Captain Bunker arrived in Port Jackson in command of the convict transport *William and Anne*, which he later used to pioneer Australia's whaling industry. He died at Collingwood in 1836. Dr Throsby built his house near the boundary of his neighbour, Richard Guise, who has the doubtful distinction of being the first person buried in the old Liverpool Cemetery now converted into Pioneer Memorial Park. Graves of earlier dates have been transferred there from elsewhere. A well-known pioneer connected with Casula was Dr John Dunmore Lang (1799-1878), the founder of the Presbyterian Church in Australia. His brother lived in a brick cottage in Liverpool and Lane made the first of a number of visits to the town in 1823. Lang Road and Dunmore Crescent were named after him.

Catani *V* is a rural area sixteen kilometres east of Koo-wee-rup. It was named after engineer Carlo Catani, for his work on the Koo-wee-rup swamp drainage plans. He also surveyed mountain roads. Lake Catani is situated on the Mount Buffalo plateau.

Caulfield *V* is a suburban city about ten kilometres south-east of the centre of Melbourne. It is famous throughout Australia because of the Caulfield Cup which has been run at the local racecourse since 1879. Caulfield may have been called after Baron Caulfield of Ireland or John Caulfield, a builder who arrived in Melbourne in 1837. Caulfield was created a district in 1857 and became a shire in 1871. It was declared a town in 1901 and a city in 1913.

Cavan *SA* is a suburb in the corporate city of Enfield in the Adelaide metropolitan area. Cavan was named after the old Earle of Cavan Hotel, which in turn may have been named after the Earl of Cavan. Cavan is a town in Northern Ireland.

Cavendish *V* is a rural township twenty-five kilometres north of Hamilton by the Wannon River. It was named after Lord Frederick Cavendish, secretary to the British Treasury.

Caversham *WA* is a locality which derived its name from the property acquired by Peter Shadwell in 1830. The property was purchased by Doctor Richard Hinds in 1855. Either one of these owners may have named the property.

Cecil Park *NSW*, a suburb of Liverpool, was originally the site of a property granted to John Wylde (1781-1859) who arrived in the colony in 1816 to take up the position of Deputy Judge Advocate. He named his property 'Cecil Hills' which eventually gave rise to the name Cecil Park. A neighbour was Barron Field (1786-1846), a judge noted for his poetry and other literary work who called his property 'Hinchinbrook'.

Ceduna *SA* has appeared on Lands Department plans since 1867 as Ceduna Plain and Ceduna Hut and Well on 'Athena Station', and this was apparently the source of the name which was the Aboriginal name for the adjoining rock holes. Baudin gave the name Murat Bay in 1802 and the area was generally known as Murat Bay Landing until 1922 when the name of the local post office was changed from Murat Bay to Ceduna.

Centennial Park *NSW* is a locality in the City of Sydney. It was declared open in 1888 to mark Sydney's centenary.

Cessnock *NSW* was named, after Cessnock Castle in Scotland, by John Campbell who was given land there in 1826.

Chadstone *V* is a suburb of the City of Waverley in the Melbourne metropolitan area. Chadstone comes from Chadstone Road and dates from about 1912. The original St Chad's Stone Church was located north of the Malvern Hills in Staffordshire in England. It was built in the seventh century by King Wulf after St Chad told him to do so to pay for the sin of murdering his two Christian sons.

Chambers Pillar *SA* was named by J. McDouall Stuart in honour of his supporter, James Chambers, in the following note written on 6 April 1860: 'It is a pillar of sandstone, standing on a hill upwards of one hundred feet in height. From the base to the top of the pillar it is one hundred and five feet, twenty feet wide by ten feet deep, and quite perpendicular, with two peaks on the top. This I have named Chamber's [sic] Pillar, after James Chambers, Esq, of Adelaide, one of the promoters of all my expeditions.' The Aboriginal name was 'Idracowra,' 'Etirkaura' or 'Itirkauwurra' meaning 'place of the adulterous male'.

Channel Country *Q* gets its name from the network of channels that criss-cross the floodplains in areas where the average rainfall is only 180 millimetres a year. But high summer rainfall in the surrounding area produces floods on the plains.

Chapman *ACT* commemorates Sir Austin Chapman (1864-1926) who became government whip when Federal Parliament first met in 1901.

Charing Cross *NSW* was so named by the Waverley Municipal Council at its meeting on 6 September 1859. No reason was given by the Council but there is not much doubt that it was named after Charing Cross in London.

Charleville *Q* is a town on the Warrego River, about 780 kilometres west of Brisbane. The explorer Edmund Kennedy passed through Charleville in 1847. W.A. Tully named the town after Charleville in Ireland when he surveyed the town in 1868. The first regular air service of Qantas began at Charleville in 1922.

Charlotte Pass *NSW*, now a ski resort near Mount Kosciusko, was named after Charlotte Adams, the then Surveyor-General's daughter, who was the first white woman to climb to the top of Mount Kosciusko.

Charlton *V*, a township by the Avoca River, 105 kilometres north-west of Bendigo, took its name from that of a pastoral station, 'Charlton West', of which it formed part. The pastoral station was taken up by W.M. Bell about 1848 on behalf of Robert Cay and William Kaye and the name, Charlton, appears on a plan made at that time by W.S. Urquhart, surveyor. The name was probably taken from Charlton, near Woolwich, England.

Charnwood *ACT* is named after the Charnwood Homestead, a former building in the Belconnen district.

Charters Towers *Q* is a city about 128 kilometres south-west of Townsville. Jupiter Mosman, an Aboriginal boy, is credited with the discovery of gold in 1872. The find led to a rush. The town became a municipality in 1877. It was named after W.E. Charters, the first mining warden. The Towers was added because of the similarity of the surrounding countryside to the Dartmoor tors in England.

Chatham *V* is a suburb of the City of Camberwell in the Melbourne metropolitan area. Chatham was probably so named because it was next to Canterbury. Chatham in England is a military and naval station in Kent.

Chatswood *NSW* is the main shopping centre of the

Municipality of Willoughby on Sydney's North Shore. An early settler, Richard Hayes Harnett, named the area Chatsworth after his wife Chattie. The name was not legally changed to Chatswood until 1895.

Chauncy Vale *T*, a wildlife sanctuary, was the home of Nan Chauncy, a well-known children's writer who used the local environment in her most famous book, *They Found A Cave*.

Chelsea *V* is a city on Port Phillip Bay made up of five suburbs—Aspendale, Edithvale, Chelsea, Bonbeach and Carrum. The first land sales in the area took place in 1865. Carrum became a borough in 1920 and the City of Chelsea was proclaimed in 1929. The city shield, like the name, is adapted from that of Chelsea in England.

Cheltenham *NSW*, a locality in Hornsby Shire on the North Shore of Sydney, received its name from the English birthplace of William Chorley, a city tailor who bought a large area of land in present day Cheltenham in 1890. His name is remembered in Chorley Road.

Cheltenham *SA* is a suburb of Woodville, a corporate city in the Adelaide metropolitan area and was named by the original owner, John Denman, who came from Cheltenham in Gloucestershire. He was a mason by trade but a farmer here. The township was surveyed and offered for sale in 1849, but for many years it remained very small with only about twenty wood or pisé cottages.

Chermside *Q* The Brisbane suburb was named after Sir Herbert Charles Chermside, Governor of Queensland from 1902 to 1904.

Chester Hill *NSW* is a suburb of Bankstown, a city located in the western part of the Sydney metropolitan area. Chester Hill was named by Miss H.A. McMillan. Two other suggested names—Hillcrest after an estate near Regents Park and Hillchester, after an English town—were rejected.

Chetwynd *V* is a rural township thirty-four kilometres south of Edenhope by the Chetwynd River. It was named in memory of Granville Chetwynd Stapylton, the deputy leader of Major T.L. Mitchell's expedition from Sydney to Port Phillip in 1835-36.

Chewton *V* is an old mining township east of Castlemaine. It was named after Lady Chewton, friend of Lady Castlemaine.

Cheyne Island *WA* is named after George McCartney Cheyne. The island bears as many natural features as the occupations adopted by Cheyne. He arrived in Fremantle in 1831, and was successively a farmer, whaling promoter, timberman, merchant, ship chandler and grazier. The features named after him include an inlet, point, creek, ledge, head and beach.

Chidley Point Reserve *WA*, located in Mosman Park, a town in the Perth metropolitan area, is on land granted in 1841 to Frederick Chidley Irwin, who administered the colony during Governor Stirling's absence and also before the arrival of Governor Fitzgerald. He was Commandant until 1852.

Chidlow *WA* was originally named Chidlow's Well after a small waterhole on the old Northam road. The watering place had been known to travellers for many years and appears to have been named after a pioneer family of Northam. Settlement began in 1883 when it became known that Chidlow's Well was to be the terminus of the second section of the 'Eastern Railway'.

Chifley *ACT* is called after Joseph Benedict Chifley (1885-1951), a Labor prime minister of Australia from 1945 to 1949.

Childers *Q* is a town in the Wide Bay district about fifty kilometres south of Bundaberg. The town was named after Hugh Culling Eardley Childers, a Melbourne politician and later a member of the British House of Commons and a Lord of the Admiralty.

Childers *V* is a rural area south of Trafalgar. It was named after Hugh Culling Eardley Childers.

Childers Cove *V* is located between Peterborough and Warrnambool. Nine children drowned there in January 1839 in the wreck of the ship *Children*.

Chiltern *V* is an old goldmining township thirty-six kilometres north-east of Wangaratta. It was named after Chiltern in England. It was previously known as Black Dog Creek, then New Ballarat.

Chilwell *V* is part of the City of Newtown and Chilwell, Geelong. Chilwell is the name of a place six kilometres from Nottingham, England. Mrs Austin, wife of Geelong landowner and mayor, James Austin, came from Chilwell and suggested her husband so name his land.

Chinamans Beach *NSW*, a locality in Mosman on Sydney's North Shore, gets its name from the Chinese market gardens which once existed behind it. Sands Directory of 1890 lists Ah Sue, a market gardener, as one of the five residents in Spit

Road. He was possibly the first of the Chinese to work vegetable gardens in the area behind the foreshore that became Chinamans Beach.

Chinchilla *Q* is a town 322 kilometres west of Brisbane on the Warrego Highway. The explorer Ludwig Leichhardt called the area Chinchilla in 1847 after *jinchilla*, the local Aboriginal name for the Cypress pines that still grow in the region today.

Chippendale *NSW*, a Sydney suburb, is named after William Chippendale, one of the earliest settlers to obtain a grant in South Sydney.

Chipping Norton *NSW*, a suburb of Liverpool, embraces part of the Moorebank Estate and a grant made to Thomas Rowley who is remembered by Rowleys Point Road. When W.A. Long owned the estate he built private racecourses there in the 1880s. His home is still standing. The Government purchased the area and subdivided it into small farm blocks for soldier settlement after World War I. Residential development has taken place there at an accelerating pace since World War II.

Chisholm *ACT* honours Caroline Chisholm (1808-1877), a philanthropist and social worker who helped over 3 000 emigrants with shelter and travel from 1849 to 1862.

Chiswick *NSW* is a suburb of the Sydney municipality of Drummoyne. It is named after the London suburb on the River Thames.

Chowder Bay *NSW*, a locality in Mosman on Sydney's North Shore, came from American whalers, who anchored there in whaling ships over a century ago. The sailors made clam chowder from rock oysters that were plentiful on the rocky shores of the bays and coves in Sydney Harbour.

Christies Beach *SA* is named after Mr L.F. Christie and Mrs R. Christie, a pioneer couple that once owned the area.

Christmas Island is an Australian territory in the Indian Ocean 2 600 kilometres north-west of Perth. When Captain William Mynors, a merchant homeward bound from the East Indies on the vessel *Royal Mary*, sighted the island on Christmas Day in 1643, the choice of name was obvious. How was he to know that Captain Cook would spot an island in the Pacific also on a Christmas Day, and for the same reason come up with the same name to confuse latter-day postal services? (To help the postman find the right Christmas Island today, the right ocean must be specified in the address).

Chullora *NSW* is a suburb of Bankstown and Strathfield in the western part of the Sydney metropolitan area. It is an Aboriginal word meaning 'native flower'.

Church Point *NSW*, a Warringah Shire locality opposite Scotland Island, got its name from a weatherboard church built there in 1872 and demolished, because of white ants, in 1932.

Churchill *V* is a township ten kilometres south of Morwell created for State Electricity Commission employees. It was named after Sir Winston Churchill.

Churchill Island *V* is located at the eastern end of Phillip Island. It was named by Lieutenant James Grant after Thomas Churchill, who had supplied numerous seeds to Grant in England and which were planted here in 1801.

Churchlands *WA* was acquired in September 1891 by the Roman Catholic Bishop of Perth, the Right Reverend Matthew Gibney. Gibney was far-sighted in his acquisition of land for the Church, and this area became known as Churchlands as a result of it being owned by the Church.

Churchman Brook Reservoir *WA* in the City of Armadale on the outskirts of the Perth metropolitan area, recalls Charles Blisset Churchman who arrived in the Swan River Colony in 1830 and took up land in the Roleystone district in 1831.

Circular Head *T* is a massive headland which Bass and Flinders named when they sighted it in 1798. The hill on Circular Head is known locally as The Nut.

Circular Quay *NSW* is the main ferry terminal in Sydney. It takes its name from the semi-circular shape of the reclamation work carried out by the engineer George Barney in the 1850s when the harbour was extensively filled in at the mouth of the old Tank Stream.

City *ACT* is the centre of Canberra City.

City Beach *WA* appears to have gained its name because it was an area developed by the Perth City Council. This name was obviously more acceptable to the Council than the more descriptive Ocean Beach. The Council drew up plans for promenades, gardens and a semicircular bathing enclosure in 1925. The first sale of land in the estate took place on 9 February 1929 and the City Council has continued developing the area up to the present date.

Clare *SA* is the centre of a rich farming and winegrowing district. Edward Gleeson and John Maynard were the first settlers in the area in 1842. Gleeson named the district after his birthplace, County Clare in Ireland.

Claremont *WA* is a town on the Swan River in the Perth metropolitan area. The name, Claremont, probably comes from a Mr Morrison, who subdivided the land he owned around the railway siding in about 1882. He called the estate 'Claremont' after his wife, Clara. The 'mont' referred to the hilly nature of the estate. Local residents disliked the name Butler's Swamp, which applied to the railway siding, and they pressed for the name to be changed to Claremont, and this was done when a new railway station was built in 1886.

Clarence *T*, a municipality on the eastern shore of the Derwent river, takes its name from Clarence Plains (now called Rokeby). Clarence Plains was named in 1793 by Captain John Hayes after his ship, the *Duke of Clarence*.

Clarence River *NSW* The discovery and naming of this river are somewhat involved. Matthew Flinders anchored at the mouth in 1799 and named it Shoal Bay, but the river was first discovered when it was crossed by four escaped convicts in 1825. For some time it was known simply as the Big River. John Oxley had discovered and named the Tweed River, but when Captain H.J. Rous of the *Rainbow* sailed into the Tweed in 1828 he believed that he was the discoverer and named it the Clarence, after the Duke of Clarence. When it later became clear that he was not the discoverer of the river, the name Clarence was transferred to what was previously known as the Big River. The Aboriginal name of the Clarence was Booroogarrabowya-neyand, 'head of the tide' (*booroogarra*, 'salt water'; *bowyra*, 'head of a creek'; *neyand*, 'top').

Clareville *NSW*, a locality in Warringah Shire, was named by Mr Stokes who built boats there in the early days of the colony.

Clarke Island *NSW* in Sydney Harbour is named after Lieutenant Ralph Clarke, an officer of the Marines who arrived with the First Fleet. He set up a garden on the island in 1790 and visited it regularly.

Clarkes Point *NSW*, a locality in Woolwich in the Municipality of Hunter's Hill, is named after John Clarke who bought land there in 1834.

Clearview *SA* is a suburb in the City of Enfield in the Adelaide metropolitan area. Clearview is a descriptive name of the locality. Like its neighbour, Broadview, it originally commanded a fine view of the Adelaide Plains and the River Torrens.

Clemton Park *NSW* is a suburb of the Sydney municipality of Canterbury. It received its name from the fact that Mr F.M. Clements, a successful businessman associated with Clements Tonic, had owned property there.

Clermont *Q* is a town about 100 kilometres west of Brisbane. Charles Gregory surveyed Clermont in 1862. He named it after the birthplace in France of Oscar de Satge, a local pastoralist and state Member of Parliament for Clermont.

Cleveland *Q* is a locality on the outskirts of the Brisbane metropolitan area, and was named by Surveyor J. Warner in 1840 after the Duke of Cleveland.

Clifton Gardens *NSW*, a locality in Mosman on Sydney's North Shore, stands on the site of a farm owned by Captain Cliffe. Today only the stone pillars and iron gates, at the entrance to what is now a block of town houses, remain from Cliffe's estate.

Cloncurry *Q* is a town about 120 kilometres east of the mining centre of Mount Isa. It was named by Robert O'Hara Burke in honour of his cousin, Lady Cloncurry of St Clerans, County Galway, Ireland.

Clontarf *NSW*, a suburb of Manly, began as one of the pleasure gardens situated around Sydney Harbour. It was named after the pleasure resort on the Bay of Dublin. Clontarf was known to the Aboriginals as Warringa, a name now used in Warringah Shire. In the middle of the beach, between Grotto Point and Clontarf Point, is Castle Rock, so named because it is thought to look like a castle. Clontarf made world headlines on 12 March 1868, when an Irishman named Henry James O'Farrell, a supporter of the anti-royalist Fenian Society, tried to kill the twenty-three-year-old Prince Alfred, Duke of Edinburgh, the second son of Queen Victoria, who was visiting Australia during a world tour on HMS *Galatea*. The Prince staggered and fell crying 'Good God, I am shot; my back is broken'. O'Farrell was seized before he could fire again. He was executed at Darlinghurst Gaol on 21 April 1868. The Prince made a good recovery and continued his journey.

Clovelly *NSW*, a locality in the Municipality of Randwick, is named after the coastal town of the same name in Devon. The first land grantee, William Charles Greville, and another settler, James Holmes—whose grant was made in 1838—had houses standing there in 1841.

Clunes *V* is an old goldmining township thirty-five kilometres north of Ballarat. Gold was discovered there in 1850. It is named after a farm in Inverness, Scotland.

Clyde *NSW* After the Clyde River in Scotland. C.M.G. Eddy, Railway Commissioner, wrote of the New South Wales Clyde: 'New Glasgow is close by, and as Old Glasgow is watered by the Clyde (to which the Duck River may be likened), perhaps "Clyde" would not be unacceptable.'

Clyde *V* is a township south-east of Cranbourne. It was named after the River Clyde in Scotland.

Coal and Candle Creek *NSW* The origin of this name will probably never be known. Two versions are current: 1. It is a corruption of Colin Campbell (Sir Colin Campbell, later Lord Clyde), who commanded the Highland Brigade in the Crimean War, and who later suppressed the Indian Mutiny.
2. It is a corruption of the Aboriginal name, Kolaan Kandahl. The latter theory is probably correct, the former the more romantic.

Coalcliff *NSW* got its name after three survivors of a wreck in 1797 set out to walk to Sydney. At this point they discovered coal and lit a fire to warm themselves. After they were rescued they reported the presence of coal, and Governor Hunter sent George Bass to investigate. Bass found several seams that extended for some distance and conjectured that they might extend throughout the range.

Coatesville *V* is a suburb of Moorabbin, a city in the Melbourne metropolitan area. It is named after Councillor L.R. Coates who served several terms as mayor.

Cobar *NSW* probably comes from the Aboriginal word *coburra*, the reddish coloured earth used by Aboriginals to paint their bodies in preparation for corroborees.

Cobden *V*, a township forty-five kilometres west of Colac, was named after the British statesman, Richard Cobden, noted for his campaign for free trade. It was formerly called Lovely Banks.

Cobram *V*, a township by the Murray River thirty-five kilometres west of Yarrawonga, comes from an Aboriginal word meaning 'head'.

Coburg *V* is an industrial and suburban city about ten kilometres north of Melbourne. Coburg is named after the Duke of Edinburgh who was also Duke of Saxe-Coburg and Gotha and who visited Melbourne in 1868. Coburg became a town in 1912 and a city in 1922. It was earlier known as Pentridge.

Coburg Peninsula *NT* P.P. King named the peninsula in 1818 during his survey of the northern coast after Prince Leopold of Saxe-Coburg, son-in-law of George IV, and

Visual Fiction H...
.30 Close.
Music. 8.30 T.O.E.

★★★★Don't miss

Have you had ...
in the last 3 mon...
aged between 16 an...
BLOOD BANK would li...
to help prepare ... ZOSTER
find out more by ringing

uncle of Queen Victoria. He later became King of Belgium.

Cockaleechie *SA* is named after the soup of that name.

Cockatoo *V* is a rural area in the foothills of the Dandenong Ranges fifty-six kilometres east of Melbourne. It was named by gold-diggers in 1859 because of the prevalence of cockatoos. It was severely burned in the 1983 bushfires.

Cockatoo Island *NSW* was officially proclaimed as a penal establishment in New South Wales by Governor Gipps in June 1841. The Aboriginals called it Biloela.

Cockburn *WA* is a city on the shores of Cockburn Sound south of Fremantle. Cockburn comes from Cockburn Sound, which was named in honour of Vice-Admiral Sir George Cockburn, an Admiralty commissioner.

Cockle Creek *NSW* was named by Colonel Paterson because of the numbers of cockles growing there.

Cocos Islands are an external territory of Australia in the Indian Ocean about 2770 kilometres north-west of Perth. They were discovered by Captain William Keeling of the East India Company in 1609. In 1827 John Clunies Ross brought a number of Malays and built up a thriving trade exporting copra and coconut to Singapore. They are also known as the Cocos (Keeling) Islands.

Codrington *V* is a rural area twenty-seven kilometres north-west of Port Fairy. It was named after Captain Codrington, master of the *Orion*.

Coffin Bay *SA* was named by Matthew Flinders in 1802 after Vice-Admiral Sir Isaac Coffin, the resident Naval Commissioner at Sheerness, in England, who assisted Flinders in the preparation of the *Investigator*. Point Sir Isaac is at the entrance to Coffin Bay on what is now Coffin Bay Peninsula.

Coffs Harbour *NSW* is named after John Korff, who is reputed to have discovered the harbour and who began cedar-getting there in 1850. The site for the village was gazetted in 1861.

Cohuna *V*, a township located south-east of Kerang by Gunbower Creek, is an Aboriginal name meaning 'brolga' (native companion), or 'camping place'.

Colac *V*, a city seventy-five kilometres south-west of Geelong, comes from an Aboriginal word *kolak* meaning 'fresh water'. The first settlement of the district was made in 1837 by Hugh Murray, a pastoralist after whom the main street of the city is named.

Colah, Mount *see* **Mount Colah**

Coleraine *V* is a township thirty-four kilometres north-west of Hamilton by Bryant Creek. The name was adopted by Surveyor Bryant because of associations with Coleraine in Ireland.

Coles Bay *T* was named in honour of Silas Cole, an early settler in the Swansea district.

Collarenebri *NSW* An Aboriginal name meaning 'place of many flowers'. In 1867 the town was gazetted as Collarindabri, and is still sometimes spelt Colarendabri, but is known locally as Colly.

Collaroy *NSW* is a suburb of Warringah Shire in Sydney's northern beaches area. Collaroy is an Aboriginal word meaning 'long swamp reeds', or 'junction of creeks'. There is also a Collaroy in North Queensland. The Warringah Shire suburb gets its name because the paddle steamer *Collaroy* was stranded on the coast there on 20 January 1881. It remained stuck until September 1884, when it was finally pulled off. One man was killed when a cable broke.

Collector *NSW* It was recorded in T.L. Mitchell's records that the Aboriginal name was Colegdar. The transition to the English word can readily be understood.

Collie *WA* is named after Dr Alexander Collie (1793-1835), a naturalist and surgeon who arrived in Western Australia in 1829. He explored the area south of the Swan River with Lieutenant Preston. The two men discovered the Collie and Preston rivers which were named in their honour by Governor Stirling. Collie town-site was gazetted in 1897.

Collingwood *V* is a city in the Melbourne metropolitan area. It was named in 1842 by Robert Hoddle at La Trobe's request after Admiral Lord Collingwood. It was originally known as Newtown. Local government began in 1855 and Collingwood became a city in 1876.

Collins Beach *NSW*, a locality in Manly, is named after Captain David Collins, Judge Advocate on Governor Phillip's staff. Collins Beach is thought to be the place where Governor Phillip was speared through the shoulder by Wilee-ma-rin, a local Aboriginal, one of a group who were feasting on a whale caught in the harbour in September 1790. Lieutenant Waterhouse managed to break off the barb which was sticking out of Phillip's back. The Governor was then rowed to Sydney Cove where the spear was removed. A plaque unveiled at Collins Beach in 1933 records the incident. There is also an Aboriginal rock carving in the area where the local historical society has erected the monument.

Collinsville *Q* is a coalmining centre eighty kilometres south-west of Bowen. It was named by the Railways Department after Mr Charles Collins, a local member of the Queensland parliament for many years.

Colyton *NSW* is a locality in the City of Blacktown on the western outskirts of Sydney. It was named by William Cox (1764-1837), military officer, roadmaker and builder who received a grant of land on 17 August 1819 on the southern side of the Western Highway. He named his grant Colyton in honour of his wife's old home town in Devon, England.

Come-by-Chance *NSW* The name was given by George and William Colless when, in 1862, they found to their surprise that they were able to purchase a sheep station in this district which is about fifty-seven kilometres south-east of Walgett.

Como *NSW* in Sutherland Shire, was named after Lake Como in Italy, by James Murphy, one-time manager of the Holt Estate and owner of the Como Pleasure Grounds, c.1888, because of similar scenery to that in Italy.

Como *WA* was developed in 1905 under the name of Como Estate. It is uncertain whether the name was selected because the land was owned by Edmund Hugh Comer or if the estate was simply named directly after Como in Northern Italy.

Concord *NSW*, a municipality in the Sydney metropolitan area, is said to have been named after Concord in Massachussetts by Major George Grose when conferring ten land grants in 1793. Grose had been a junior officer in the British forces during the American War of Independence in 1776 and Concord was the site of the skirmish that began the war. The 1795 settlement, made east of the present Concord station, was occupied by both civil and military settlers which was unusual at that time in the colony's history.

Condell Park *NSW* is a suburb of Bankstown, a city located in the western part of the Sydney metropolitan area. Condell Park was named after Mr Ousley Condell, who had four grants of land in the area.

Condobolin *NSW* is an Aboriginal name meaning 'hop bush'. William Lee established a run which he called 'Condoublin' some time before 1840.

Coober Pedy *SA* is an Aboriginal word said to mean 'white fellow's hole in the ground', in reference to the underground dwellings of the opal seekers. It was originally Stuart's

Range, but postal authorities had it changed due to confusion with Stewart's Range in the South East about 1920. Stuart Range in the adjacent range of hills was named after the explorer, John McDouall Stuart.

Coochie Mudlo *Q* is Aboriginal for 'red stone'.

Coogee *NSW*, a beachside locality in the Municipality of Randwick is supposed, according to popular tradition, to come from an Aboriginal word meaning 'bad smell' or 'stink', perhaps because of the rotting seaweed. However, like many popular name derivations, this is suspect.

Coogee *WA* takes its name from nearby Lake Coogee. Originally this lake was named Lake Munster after Prince William, the Earl of Munster, and later King William IV. The Aboriginal name Kou-gee was recorded in 1841 by Thomas Watson and, variously spelt, gradually gained pre-eminence over the old name. It is possible that the spelling used for the locality may have been influenced by that used for a beach resort in New South Wales although the Western Australian pronunciation differs greatly from the short vowel sound used in the east.

Cook *ACT* is named after Captain James Cook (1728-1779), the navigator who charted the east coast of Australia in 1770. It also recalls Sir Joseph Cook (1860-1947) who was Prime Minister of Australia from 1913 to 1914.

Cooks Hill *NSW* is named after Tom Cook, a wealthy squatter who lived in a cottage at the corner of Auckland and Laman Streets, Newcastle.

Cooktown *Q* is a small tourist centre located at the mouth of the Endeavour River, about 170 kilometres north of Cairns. In June 1770 Captain James Cook beached his ship, the *Endeavour*, for repairs and named the Endeavour River. The town that developed later on the site was named in Cook's honour.

Coolangatta *Q* was named after a schooner that was wrecked off the coastline in 1846.

Coolbellup *WA* is located on land taken up by George Robb in 1830 and takes its name from the Aboriginal name for the lake near the eastern end of Robb's grant.

Coolgardie *WA* is said to have come from an Aboriginal word Coolacaaby which describes the vegetation and waterholes found in the area. Gold was discovered there in 1892.

Coolum *Q* is a seaside resort 134 kilometres north of Brisbane. The name is derived from an Aboriginal word *gulum* or *kulum,* which means 'without', 'wanting'. This refers to Mount Coolum (208m) which has no peak.

Cooma *NSW* comes from a contraction of Coombah, an Aboriginal word meaning 'big lake' or 'open country'. Other meanings assigned to it are 'swamp', and 'junction of two streams'.

Coonabarabran *NSW* is an Aboriginal word sometimes wrongly translated as 'an inquisitive person'. In fact it means 'a peculiar smell', and was named by the Aboriginals because of the odour of the river weed which grows there. In the form 'Coolabarabyan' it was the name of a station owned by James Weston in 1848. The town was first surveyed in 1859.

Coonalpyn Downs *SA* is located in the upper south-east of the state. Its name comes from an Aboriginal word meaning 'barren woman'. It was called the Ninety Mile Desert until 1911. John Barton Hack, who arrived in Adelaide in 1837, had a pastoral run by the name of 'Coonalpyn'. In 1949, after considering objections to the name by several branches of the Agricultural Bureau, Cabinet decided the name Coonalpyn Downs was to stand. Objectors desired the name Ninety Mile Downs. The town is known only as Coonalpyn.

Coonamble *NSW* is an Aboriginal word meaning 'lot of dirt'.

Coonawarra *SA* was founded as a fruit colony by John Riddoch about 1895. The name is of Aboriginal origin but its meaning is disputed. The anthropologist, Dr N.B. Tindale, says it derives from *kuneia-warama* meaning 'to light a fire'. Both A.W. Reed and Rodney Cockburn say it means 'honeysuckle rise'. A note in the South Australian Lands Department records taken from *Railway Nomenclature* (1915) says that it comes from an Aboriginal word *coon* meaning 'big lip' and *warra* meaning 'house'. According to this note, the name was applied by a local Aboriginal to a house lived in by a man with a remarkably big lip. This may have referred to John Riddoch.

Cooper Creek *Q SA* was discovered by Charles Sturt on his journey to the Simpson Desert in 1845. It was named in honour of Sir Charles Cooper, the first South Australian Chief Justice. The Aboriginal name was Barcoo. T.L. Mitchell crossed it in 1846 and called it Victoria River. Other names that had applied to various parts of the stream were Strzelecki and Kennedy. Edward B. Kennedy, who was Mitchell's assistant, traced the creek from its source, the Thompson River and the Barcoo in Queensland. The explorers, Burke and Wills, died on the Cooper near what is now Innamincka.

Coorong *SA* is a series of narrow lagoons separated from the sea by the Young-husband Peninsula. It is derived from an Aboriginal word, *kurangh* meaning 'narrow neck'. The Coorong was first seen by Strangways and Hutchinson in 1837. It was explored by W.J.S. Pullen and Dr R. Penny in 1840 when investigating the reported wreck of the *Maria* and the murder by Aboriginals of her passengers. The name was officially reported to Major O'Halloran in 1840.

Coorparoo *Q* is a locality in the Brisbane metropolitan area and was named by local residents at a meeting on 22 March 1875. It is an Aboriginal name meaning 'gentle dove'.

Cootamundra *NSW* came from an Aboriginal word meaning 'turtles', 'swamp' or 'low lying'. In the form 'Cootamandra' it was a stock station owned by John Hurley in the 1830s. Later it was known as the village of Cootamundry. The first town residential lots went on sale in 1862.

Coot-tha, Mount *see* **Mount Coot-tha**

Cope Cope *V* is a farming area twelve kilometres south-east of Donald. It comes from an Aboriginal word *gope gope*, meaning 'large lake into which smaller lakes empty'.

Coraki *NSW* is a corruption of an Aboriginal word meaning 'blowing up the mountain'. Other renderings are 'plain turkey', or from *kurrachee*, 'mouth of river'.

Corangamite, Lake *see* **Lake Corangamite**

Corio *V* is a shire created on 21 June 1864. It is also the name of an industrial suburb within the shire. The Aboriginal word, *coraiyo*, meant 'small marsupial' or 'sandy cliffs' (i.e. of Corio Bay). The district was once known as Cowie Creek after the first settler. Corio is now part of Greater Geelong.

Corner Inlet *V* is a large bay north of Wilsons Promontory. It was discovered and named by Goerge Bass in 1798. It is a descriptive name.

Coromandel Valley *SA* takes its name from the ship *Coromandel* which arrived on 12 January 1837 under the command of Captain William Chesser (a street in the City of Adelaide is named after him). Ten of the crew deserted and hid in the locality until the ship sailed.

Coromby *V* is a rural area north-east of Murtoa. It is an Aboriginal word meaning 'big waterhole'.

Corowa *NSW* comes from the Aboriginal word *corowa* or *currawa* which is the Curra pine which yielded gum used by the Aboriginals to fasten the heads of spears to the shafts. Another translation is 'rocky river'.

Corrimal *NSW* comes from the Aboriginal name of the adjacent hill which was Kori-mul.

Corryong *V*, a township 120 kilometres east of Wodonga, is a corruption of *cooyong*, an Aboriginal name for the bandicoot. It was formerly called Gravel Plains.

Cosgrove *V* is a rural area twenty-four kilometres east of Shepparton. It was named after an early settler, Peter Cosgrove. It was originally called Pine Lodge East.

Costerfield *V* is an old goldmining area eleven kilometres north-east of Heathcote. It was named after two miners, James Coster and Edwin Field, who discovered gold there in 1862.

Cotteril, Mount *see* **Mount Cotteril**

Cottesloe *WA* is a town six kilometres from Fremantle and twelve kilometres from Perth. Governor Broome chose the name Cottesloe in 1886, and declared it a suburban area. It was named from the title of the first Baron Cottesloe (pronounced 'Cotslow') of Swanbourne, Buckinghamshire, England.

Cotton, Mount *see* **Mount Cotton**

Cowan *NSW*, a locality in Hornsby shire on the North shore of Sydney, is probably named after the place in Scotland with the same name, althogh some reference books give it as an Aboriginal word for 'big water' or 'uncle'.

Cowes *V* is a township on Phillip Island. It was named by Commander Henry Laird Cox in 1865 when surveying Western Port. He named it after the well known port on the Isle of Wight.

Cowra *NSW* is an Aboriginal word meaning 'rocks'.

Cox's Creek *SA* takes its name from Robert Cock. According to one authority, H.C. Talbot, Cock led a party of five men who set out from Adelaide in June 1838 in an attempt to find a passable track from the hills around Crafers to Mt Barker. However, Rodney Cockburn has recorded a different story supplied by J. Dunn, who erected the mill at Bridgewater. According to this story, Cock used to go from Balhannah to Adelaide for supplies once a fortnight and was stuck for ten to twelve days in the creek that now bears his name near a wayside hotel called "The Deanery' run by a Mr Dean. However, both sources agree that Cock's Creek was later corrupted to Cox's Creek.

Cox's Pass; Cox's River *NSW* are named after William Cox, who was appointed Superintendent of Works in 1814 and made history by constructing 160 kilometres of road over the

Blue Mountains in six months. Governor Macquarie, who traversed the road in April and May 1815, expressed his approval by naming the pass over Mount York and the river beyond the summit (where Divine Service was held) after the celebrated roadmaker.

Crackenback Range *NSW* rises above Thredbo alpine village. The name Crackenback is said to be a pidgin English word used by the Aboriginals. Revd W.B. Clarke's *Researches in the Southern Goldfields of New South Wales* recorded on his journey over the Southern Alps in 1851, says 'Crackemback (Crack-em-back) the origin of the name I was told by the Aborigines who speak English to be the steepness of the ranges, to ascend which would "crack a man's back".'

Cradle Mountain *T* is used because its shape resembles the cradle which miners use to wash gold from gravel.

Crafers *SA* is named after David Crafer who started a hotel there in March 1839. The hotel was the first one licensed outside what is now known as Adelaide and its suburbs. Bushrangers were later captured at Crafers after they had baled up Mrs Crafer and provided free drinks for all the customers. By the time the police arrived the bushrangers were too drunk to resist capture.

Craigieburn *V* was probably named by a group of Scots who settled in the district before 1850. The bluestone inn called 'Craigie Burns' was a stopping place for coaches.

Cranbourne *V* is a township sixteen kilometres south-east of Dandenong. It was first surveyed in 1852 by H.B. Foot and named after Viscount Cranbourne, the son of Lord Salisbury. Cranbourne was created a district in 1860 and a shire in 1868.

Cremorne *NSW*, a locality in Mosman and North Sydney, was named by Carke and Woolcott, two promoters who ran ferries from Woolloomooloo and charged the public for use of their pleasure grounds which they called Cremorne after the famous Cremorne Gardens of London. It offered a quadrille band and a huge dancing-stage. There was a merry-g-round for the children. Other attractions included skittles, archery, and rifle shooting and fireworks at night. The Cremorne amusement gardens opened on Easter Monday 1856 but the business failed six years later. In 1893 an important coal discovery at Cremorne was announced. Coal was stuck at a depth of 1000 metres but because of public opposition, the venture was abandoned.

Cressy *T* owes its name to an English syndicate, the Van Diemen's Land Establishment, which was formed in 1826 to breed horses and other stock in a grant of land stretching from 'The Hermitage' on the Liffey River to 'Cressy House', which served as the headquarters on the Lake River. The syndicate later changed its name to the Cressy Company. Cressy, or Crecy, was the historic battle place where an ancestor of the company's first manager, Captain B.B. Thomas, was knighted by King Edward III. The township of Cressy dates from 1855.

Cressy *V* is a rural township thirty-eight kilometres north of Colac. A Frenchman and his wife kept an inn here in 1840 and a French name was therefore given by the settlers. The Aboriginal name was Bitup.

Creswick *V*, a township sixteen kilometres north of Ballarat, is named after John and Henry Creswick, two early settlers in the district in 1842. The township was first known as Creswick Creek. The Aboriginal name was Collum-been.

Crib Point *V* is a township eight kilometres south of Hastings on the west coast of Western Port. It was so named because two of the settlers, R. and J.Hann, built a crib or hut here.

Cromer *NSW* is a suburb of Warringah Shire. Cromer gets its name from Cromer Golf Club, which in turn is named after Cromer in Kent, because the bird life from nearby Narrabeen lagoon reminded a lot of people of the birds in Cromer in England.

Cronulla *NSW*, in Sutherland Shire, is thought to be a European variation of the Aboriginal word Kurranulla, 'a place of pink shells'.

Crookwell *NSW* is thought to be named after the Crookwell River which was first mentioned in 1828 by Surveyor Dixon when marking out a road 'to Kyama on Crookwell Road'. There does not appear to be a town of this name elsewhere and it may have originated from a local property. When land was reserved here in the 1830s it was called Kiama or Kiamma, a common name meaning 'good fishing ground'. The town-site was surveyed in 1860, and the name changed to that of the river, which had been known by this name for many years.

Crows Nest *NSW*, a locality in North Sydney, recalls the name of a cottage built by Edward Wollstonecraft on the grant of land he obtained on the North Shore in 1819. North Sydney Demonstration School now stands on part of the old Crows Nest farm where Alexander Berry completed a larger 'Crows Nest House' in 1850. The house itself was a little more on the hill but the school is in its old grounds.

Croydon *NSW* is a suburb in the Municipality of Burwood in the Sydney metropolitan area. Croydon was suggested by the local council in 1874 because the original name of its railway station Five Dock conflicted with a locality of that name about five kilometres north of the station. Croydon may have originated from the suburb south of London or a village in Cambridgeshire. The London suburb name means 'valley where wild saffron grew', while the Cambridgeshire Croydon means 'valley frequented by crows'. The choice of Croydon is said to have been influenced by the fact that Croydon was the same distance from the old Homebush racecourse as the London suburb of Croydon was from another racecourse. However there is no documentary proof of this story.

Croydon *SA* is a suburb of Hindmarsh in the inner metropolitan area of Adelaide. Croydon was named by Patrick Coglin, a mayor of Hindmarsh, who retained the name of 'Croydon Farm' when he cut up the area into allotments in 1855. He called it after the town in Surrey.

Croydon *V* is a suburban city at the foot of the Dandenong Ranges about thirty kilometres east of the centre of Melbourne. Croydon took its name from the English town of the same name. It was first known as White Flats. The founder of Croydon was William Turner (1808-1893), a Yorkshireman and former captain who arrived in Sydney in 1837. Later, he set up the first tailor's shop in Elizabeth Street, Melbourne, before taking up land for a cattle station at Croydon. Local government began in the area in 1856. Croydon was part of the Shire of Lilydale from 1872 until 1961. Croydon was proclaimed a city in 1971.

Croydon Park *NSW* is a suburb of the Sydney municipality of Canterbury. It received its name after the Croydon Parents and Citizens' Association sent a petition in October 1914 asking that a post office be established at Croydon Park.

Crystal Brook *SA* was discovered in 1839 by E.J. Eyre while on his third expedition, and named because of the clarity of the water. The Aboriginal name was *mercowie* meaning 'clear water'.

Cudgee *V* is a rural area twelve kilometres north-east of Warrnambool. It comes from an Aboriginal word meaning 'kangaroo skin'.

Cudlee Creek *SA* probably comes from the Aboriginal word for 'dog', *kudlee*. Wild dogs were numerous in the area. It is also argued that it was originally called Chudleigh Creek but the former is the more acceptable derivation.

Culgoa River *NSW Q* flows 113 kilometres through southern Queensland into New South Wales where it meets the Darling River. Its total length is about 320 kilometres. It was named by the explorer Sir Thomas Mitchell in 1845. *Culgoa* is an Aboriginal word probably meaning 'running through'. Another translation is 'Leatherhead', or 'Friar-bird'. The name of these honey-eaters comes from the absence of feathers on the head.

Cundletown *NSW* has a mixed Aboriginal and English name. *Cundle* is a native plant, the fruit of which resembles carrot.

Cunnamulla *Q* is a town on the Warrego River about 970 kilometres from Brisbane. The name is Aboriginal for 'big waterhole'. The 'Cunnamulla' station was established on the east side of the Warrego River by Cobb & Co.

Cunninghams Gap *Q* is the pass between the Darling Downs and Brisbane. It is named after the explorer and botanist, Allan Cunningham, who found the way from the sea at Moreton Bay to the fertile Darling Downs in 1828.

Curl Curl *NSW* is a beach suburb of Warringah Shire. Curl Curl gets its name from Aboriginal words which have not been documented.

Currumbin *Q* is an Aboriginal name for a species of pine trees.

Curtin *ACT* is named in honour of John Joseph Curtin (1885-1945), Labor Prime Minister of Australia from 1941 to 1945 during World War II.

Curtis, Port *see* **Port Curtis**

Cygnet *T* was first explored and named Port de Cygnes because of the numerous black swans in the estuary by Huon Kermandec, a French captain with Bruny D'Entrecasteaux in 1793. A convict settlement operated at Cygnet during the 1840s. The town was earlier known as Lovett after the Auditor-General at the time of settlement. It was changed to Cygnet officially in 1915.

Daceyville *NSW* is a suburb of the Sydney municipality of Botany. It was Australia's first garden suburb. The idea of garden cities came from places like Letchworth and Welwyn Garden City, in England. Daceyville was named after John Rowland Dacey, a local member of state parliament, from 1895 until his death in 1912. Dacey was the first to suggest the idea of building a model suburb in Australia. The plan was drafted by Sir John Sulman, the pioneer of town planning in this country.

Daguilar Range *Q*. Named after Sir George S. D'Aguilar, a British army officer. The name was probably conferred by T.L. Mitchell. The D'Aguilar Range in Tasmania is also doubtless named for him.

Daintree, Mount *see* **Mount Daintree**

Daintree River *Q* was discovered by G.A.F.E. Dalrymple in 1873 and named after Richard Daintree, who became Government Geologist for Northern Queensland in 1869 and then became Agent-General in London.

Dalby *Q* is a town on the fertile Darling Downs about 220 kilometres from Brisbane. Dalby was surveyed in 1854 and named after the town of the same name on the Isle of Man.

Dalkeith *WA* was assigned to Adam Armstrong on 8 September 1831. He named the property Dalkeith after his home town in Scotland and had completed his location duties by January 1839. These included the construction of a dwelling which was called 'Dalkeith Cottage'. The Aboriginal name for the place where this cottage stood was Katamboordup.

Daly River *NT* was named by Colonel B.T. Finniss, the first South Australian Government Resident of the Northern Territory in 1864. He had been Premier of South Australia and named the river after Sir Dominick Daly, Governor of South Australia.

Daly Waters *NT* was discovered by J. McDouall Stuart in 1862 and named after Governor Daly of South Australia.

Dampier *WA* is named after the nearby Dampier Archipelago, which in turn was named in 1821 by P.P. King after the English buccaneer William Dampier, who anchored there briefly in 1699.

Dandenong *V* is an industrial and residential city about thirty kilometres from the centre of Melbourne. It is also the name of a mountain, a mountain range and a creek. There is no clear record of how the name of Dandenong came into being. The honour is generally given to Captain Lonsdale, who in an official report on the district in 1837 spelt the name Dan-y-nong, this being part of the area inhabited by the tribe of Westernport Aboriginals known as Wooeewoorong. The original surveyors spelled the name Tanjenong or Tangenong, which they thought from the explanation of the local Aboriginals meant 'lofty' or 'lofty mountain'. The first white man to pioneer the district of Dandenong was Joseph

Hawdon, who established a cattle station on what is now the present site of Dandenong in 1837. Dandenong became a city in 1959.

Dangar, Mount *see* **Mount Dangar**

Danger Point *see* **Point Danger**

Dapto *NSW* is said to have been taken from the word 'taptoe' which described the way a lame Aboriginal chief walked.

Darke Peak *SA* was named after John Charles Darke, an explorer who died here. Darke led an expedition north-west of Port Lincoln in 1844, and wrote in his report that 'no land existed beyond the Gawler Ranges available to the Port Lincoln settlers'. On 23 October, on his return journey, he was wounded by the spears of hostile Aboriginals and died on the night of the 24th at the place now known as Darke Peak. An obelisk was erected over his grave by the South Australian Government in 1910. The Aboriginal name was Carrapee. The town was proclaimed as Carappee on 4 June 1914 and changed to Darke Peak on 19 September 1940.

Darling *V* is a locality in the City of Malvern in the Melbourne metropolitan area. Darling is named after Sir Charles Darling, Governor of Victoria from 1863 to 1866.

Darling Downs *Q* were named after Sir Ralph Darling, Governor of New South Wales, by Allan Cunningham, the botanist who visited the area in 1827 and found the route between the Downs and the sea in 1828.

Darling Point *NSW* is located on Sydney Harbour in the Municipality of Woollahra. Darling Point was named by Major General Ralph Darling during his governorship which lasted from 1825 to 1831. He called it Mrs Darlings Point to honour his wife but the Mrs has since been lost.

Darling Range *WA* was originally called General Darling's Range in 1827 by Governor James Stirling who named it after Governor Darling of New South Wales. It was later shortened to Darling Range.

Darling River *NSW* Named after Sir Ralph Darling, the river was discovered by Charles Sturt, who had set out to find what happened to the rivers that flowed westward. The Macquarie River was followed and great excitement was engendered by the discovery of a larger stream—balanced by disappointment when it was found that its waters were salty. The saltiness was caused by the river flowing over beds of salt deposited by an ancient inland sea. Sturt, who had received much encouragement from the Governor, conferred his name on the newly discovered river on 2 January 1829. It should be noted that in its upper course it has several names—these usually being accepted in sequence as the Severn, Dumaresq, Macintyre and Barwon. As the Darling it runs from its junction with the Culgoa to its meeting with the waters of the Murray.

Darlinghurst *NSW* is a locality in the City of Sydney. It is named after Governor Darling. The word *hurst* in Old English meant 'a wooded hill'.

Darlington *NSW*, a Sydney suburb, is named after Governor Ralph Darling. Darlington had the distinction of being the smallest self-controlled municipality in the Sydney metropolitan area when it was incorporated in 1864. By the year 1917 there was not a single vacant lot in the municipality.

Darlington *WA* is a locality in Mundaring Shire on the outskirts of the Perth metropolitan area. 'Darlington' was originally the name of Alfred Waylen's vineyard established in 1885. A railway siding there became a regular stopping place named Darlington in 1892.

Dartmoor *V* is a rural district fifty-five kilometres north-west of Portland near the junction of Glenelg and Crawford rivers. It was named after Dartmoor in Devonshire, England.

Darwin *NT* takes its name from Port Darwin, which was sighted by John Lort Stokes of HMS *Beagle* in 1839. Captain J.C. Wickham called it after his friend and former shipmate, Charles Darwin, the famous naturalist. In 1869, the South Australian Surveyor-General, George Goyder, selected Port Darwin as the site for a permanent settlement. Goyder named his new town Palmerston after the British prime minister, Viscount Palmerston, but it was usually known as Port Darwin. Palmerston remained the official name until 1911, when control of the Northern Territory was transferred from South Australia to Commonwealth administration and it reverted to the original name, Darwin.

Davidson *NSW* is a suburb of Warringah Shire. Davidson is named in memory of Sir Walter Edward Davidson, a former governor of New South Wales.

Dawes Point *NSW* is a locality in the City of Sydney. It was named after William Dawes (1792-1836), a soldier who arrived in Sydney with the First Fleet. Dawes was also a scientist. He set up the first observatory in Australia at The Rocks in 1788. The present domed building on Observatory Hill was not completed until 1858.

Dawson River *Q* was named by Ludwig Leichhardt after Mr R. Dawson of Hunter's Hill, New South Wales, a supporter of Leichhardt's exploration expeditions.

Dawsons Hill *NSW* is named after Robert Dawson, the first agent of the Australian Agricultural Company, to whom a grant was made in this area in 1828.

Dayboro *Q* was named after W.H. Day who established a sugarcane plantation using Kanaka labour.

Daydream Island *Q* was originally named West Molle in 1815 by Lieutenant Charles Jeffreys of the brig *Kangaroo*. He called it after Colonel George Molle the Lieutenant-Governor. The island was renamed for tourist publicity purposes.

Daylesford *V* is a township on Wombat Creek forty-five kilometres north-east of Ballarat. Gold was discovered there in August 1851. It was originally known as Jim Crow Creek and then as Wombat Hill. Its present name comes from Daylesford in Worcester, the home of Warren Hastings, Governor-General of India.

De Grey River *WA* was discovered in 1861 by F.T. Gregory and named in honour of the president of the Royal Geographical Society, the third Earl de Grey, Earl (and later Marquis) of Ripon.

Deadmans Gully *Q* got its name because a station hand, sent from 'Blue Hills' outstation to Calliope to warn owners of an impending raid by Aboriginals, was killed there.

Deakin *ACT* honours Alfred Deakin (1856-1919) who was three times Prime Minister of Australia in 1903-1904, 1905-1908 and 1909-1910.

Dee Why *NSW* is a suburb in Warringah Shire in Sydney's northern beaches area. The origin of its name is something of a puzzle. Surveyor James Meehan made an entry in 1815 'Dy Beach', but failed to give any further explanation. Manly-Warringah historian, Charles Mcdonald, thinks that Meehan used Dy as a surveyor's term for unfinished line, but modern surveyors claim this term was not used in Meehan's day.

Deer Park *V* is a locality in the City of Sunshine on the outskirts of Melbourne. Its name recalls the fact that the Melbourne Hunt Club kept deer in the district last century.

Delegate *NSW* Apparently a corruption of an Aboriginal word, the meaning of which has not been discovered. On an early plan it appeared as Dilligat, and a local station was named 'Deligat'.

Deloraine *T* was named by Surveyor Scott after Sir William Deloraine in *Lay of the Last Minstrel* by his kinsman, Sir Walter Scott.

Denham Court *NSW*, a suburb of Liverpool, was the name given by Judge Advocate Richard Atkins to the property that was a land grant to him. It is said to be named after his family's home in Denham England.

Deniliquin *NSW* was first visited by white men when John Webster and James MacLaurin reached the Edward River in 1841 and named it after their employer, Edward Howe. It is possible that the present name derives from Denilakoon, an Aboriginal chief and a renowned wrestler. Deniliquin can be broadly translated as 'wrestling ground'. Employees of the pioneer businessman, Benjamin Boyd, established the 'Deneliquin' run in the early 1840s. Deniliquin was gazetted a municipality in 1868.

Denistone *NSW*, a locality in the Municipality of Ryde, is named after a farm set up in 1829 by Dr Thomas Forster, who married Eliza Blaxland in 1817.

Dennis *V* is a suburb of the City of Northcote in the Melbourne metropolitan area. Its name honours Samuel Dennis, a stonemason who built a stone house in Walker Street and became a mayor of the city.

D'Entrecasteaux Channel *T* is named after the French Admiral, Bruni D'Entrecasteaux, who sailed through the channel in 1792.

Derby *T* is named after Derby in England. The Briseis Mine near Derby, named after the racehorse which won the 1876 Melbourne Cup, became well known in the mining world.

Derby *WA* was declared a town in 1883 and named after Lord Derby who was then Secretary of State for the colonies.

Derrinallum *V* is a rural district near Mount Elephant. It is an Aboriginal word meaning 'tern' or 'sea swallow'.

Derwent *T* was discovered in February 1793 by a party from the French vessel *La Recherche* under Admiral D'Entrecasteaux, who named it Riviere du Nord (Northern River). Two months later, Lieutenant John Hayes gave the river its present name after the Derwent in England.

Devil's Marbles *NT* So named on account of the granite boulders varying in size from twenty centimetres to six metres in diameter, found about one hundred kilometres south of Tennant Creek.

Devonport *T* was named when the townships of Torquay and Formby joined to form the town in 1890. The town was called after its English namesake because it was in the County of Devon.

Dharruk *NSW* is a locality in the City of Blacktown on the western outskirts of Sydney. It is the name of the local Aboriginal tribe associated with the area west of Sydney.

Diamantina River *Q* is an inland river. The explorer, John McKinlay, found the river in 1868 while searching for Burke and Wills. He named it Muellers Creek after the botanist, Ferdinand von Mueller. William Landsborough rediscovered the river in 1866. He named it Diamantina after the wife of Sir George Ferguson Bowen, the first governor of Queensland.

Diamond Bay *NSW* first appeared in a Report and Map of Harbour Defences, dated 3 January 1863. The origin of the name is not known.

Diamond Valley *V* is a shire about thirty kilometres from Melbourne. Diamond Valley takes its name from Diamond Creek, which is said to have got its name because the original surveyors found the water so clear and sparkling that they could see crystals on the bottom. According to another legend, Diamond was the name of a bullock lost in the creek.

Dianella *WA* was chosen from twenty-six names submitted by the Nomenclature Advisory Committee to the Perth Road Board in 1958. The board chose Dianella, the botanical name of a small blue lily common in the area.

Dickson *ACT* is named after Sir James Dickson (1813-1901) one of the founders of the Australian Constitution.

Diggers Creek *NSW* flows under the main road to Perisher Valley. It is named after gold-diggers who joined a rush to Crackenback at the head of the creek in 1860. A quartz vein runs along the valley to the west of Alpine View and heaps of rubble and broken quartz are still visible where the diggers operated. It was hard work with very little result. Most of the gold found was alluvial.

Diggers Rest *V* took its name from a station with rough shelters set up by Caroline Chisholm for goldminers' families journeying to the goldfields at Bendigo in the 1850s.

Dight Falls *V* is a locality in the City of Northcote in the Melbourne metropolitan area. The falls are named after John Dight who built a flour mill beside the falls in 1834.

Dimboola *V*, a township thirty-six kilometres north-west of Horsham, was originally called Nine Creeks. Its present name was given by Surveyor J.G.W. Wilmott in 1863 from Dimbulah (land of figs) in Sri Lanka.

Dingley *V* is a locality in the City of Springvale on the outskirts of Melbourne. Dingley was named in the early 1860s by the couple that had the first church built. They were Thomas and Mary Attenborough from Dingley, England.

Dinosaur Point *WA* Fossilised footprints of a dinosaur have been found at the foot of the cliffs. Dr E.H. Colbert of the American Museum of Natural History has stated that they were made by a carnivorous type of dinosaur, about eight metres long and two metres tall. Dr Colbert and Duncan Merrilees decided it was a theropod dinosaur and intended to name it as a new species and a new genus.

Diogenes, Mount *see* **Mount Diogenes**

Dirk Hartog Island *WA* is named after Dirk Hartog, a Dutch captain who, in 1616, became the first white man to step ashore on the west coast of Australia. He nailed a pewter plate to a post as a record of his visit. In 1697, the plate was replaced by one put there by Willem de Vlamingh. Hargtog's plate is now in a museum in Amsterdam. A copy is housed in a museum in Fremantle along with the Vlamingh plate.

Disappointment, Mount *see* **Mount Disappointment**

Discovery Bay *V* was first sighted on 3 December 1800. James Grant on the brig *Lady Nelson* became the first European to sight this area of coastline. However, it was not named until 1836, when T.L. Mitchell sailed down the lower reaches of the Glenelg River. On 20 August 1836 he found that the river 'terminated in a shallow basin . . . choked up with the sands of the beach . . . so our hopes of finding a port at the mouth of this fine river were at an end'.

Dobroyd Point *NSW*, a suburb of Ashfield, takes its name from Dobroyd Farm owned by the Ramsay family who first settled in the early 1800s.

Dolans Bay *NSW* in Sutherland Shire was named after Patrick Dolan who purchased land there on 17 January 1856.

Donald *V*, a township forty-two kilometres west of Charlton by Richardson River, was named after a pioneer family. It was formerly called the Bridge or Richardson Bridge.

Doncaster *V* is part of the City of Doncaster and Templestowe in the Melbourne metropolitan area. Doncaster was named after Doncaster in England by a landowner who recognized the association of Doncaster, Burnley and Richmond in England. Settlers first began arriving at Doncaster in the early 1850s. During these early years, Doncaster existed almost entirely for woodcutting and charcoal burning. Doncaster is the oldest fruit growing area in Victoria. The industry was founded by two German residents, Oswald and Gothlieb Theile, whose homestead Friedensruh still stands in Waldan Court, Doncaster. Waldan means 'clearing in the forest', and was the name given to the area by the Lutherans who settled there in the 1850s. Doncaster and Templestowe became a city in 1967.

Doo Town *T* Pamela Davis, in *People*, 22 February 1967, wrote that the town was founded about 1937 by the tradespeople of Hobart. Each of them named his house Doo—Doo Us, Doo I, Doo Me, followed by Av Ta Doo, Thistle Doo, Doo Nix, Didgeridoo, How Doo U Doo, Xanadoo, Make Doo, Doo US 2, etc. A sign was erected 'Welcome To Doo Town Doo Drive Slowly'.

Doonside *NSW* is a locality in the City of Blacktown on the western outskirts of Sydney. It was named by Robert Crawford (1799-1848) who named his grant 'Doonside' after his home town in Scotland.

Dora Creek *NSW* is supposed to come from a corruption of an Aboriginal word for 'creek'.

Dorrigo *NSW* A contraction of the Aboriginal word *dondorrigo* which is said to mean 'stringybark'.

Double Bay *NSW* is located on Sydney Harbour in the Municipality of Woollahra. Double Bay has two beaches separated by a small headland, although it is not a 'double bay' in any exact topographical sense.

Double Island Point *Q* was named by Captain Cook in May 1770 during his voyage in the *Endeavour*. He chose the name because, in his words, 'the point itself is so unequal, that it looks like two small islands lying under the land'. He added that 'it may be known by the white cliffs on the north side of it'. In 1842, Andrew Petrie explored this district and named the point 'Brown's Cape' after Brown, the mate of the *Stirling Castle,* who was killed and eaten by Aboriginals on this headland. However the name 'Double Island Point' has been retained in official records.

Doubleview *WA* was owned in 1930 by two speculative builders named Dudley and Dwyer. They chose the name because of the extensive views obtainable from the higher parts of the estate, both to the coast to the west, and hills and city to the east.

Douglas Springs *Q* is named after a Police Sergeant Douglas. Around 1850 a tribe of Aboriginals at the head of Kroombit Creek killed a shepherd. Sergeant Douglas led a posse from Banana to the site, herded the tribe into a gully and shot them. The incident became known as the Douglas Springs massacre. The gully is now Douglas Springs.

Dover Heights *NSW* is in the Sydney municipality of Waverley. Dover Heights appears to have been first mentioned in municipal records in March 1886. A road leading from Rose Bay to the heights prior to this period was called Dover Street (now Dover Road) —hence the name Dover Heights.

Doveton *V* is a locality in the City of Dandenong on the outskirts of the Melbourne metropolitan area. Doveton honours Frances C. Doveton, Commissioner of the Central Goldfields and later a warden and chief magistrate.

Downer *ACT* honours Sir John Downer (1844-1915), a former premier of South Australia who became a senator after federation in 1901.

Dromana *V* derives from either County Cook or County Waterford in Ireland.

Drouin *V* is a township eleven kilometres west of Warragul. Some authorities say that Drouin is named after a Frenchman who developed a process of extracting metals from ores. Others say it comes from an Aboriginal word *douran* meaning 'north winds' and this explanation is considered the most likely one.

Druitt, Mount *see* **Mount Druitt**

Drumcondra *V* is a suburb of Geelong. Surveyors McDonald and Garrard in 1844-45 named this subdivision by Western Beach, Geelong, after a suburb in Dublin, Eire.

Drummoyne *NSW* is a municipality in the Sydney metropolitan area. Drummoyne comes from two Gaelic words meaning 'flat-topped ridge'—*drum*, 'ridge' and *moyne*, 'flat' or 'plain'. The suburb takes its name from Drummoyne House, built in 1854 by William Wright, who constructed the railway in the Hunter Valley from Newcastle to East Maitland. On Wrights Point in front of his Drummoyne mansion, William Wright built a private landing stage and stone steps. When Wright's estate was subdivided in the 1880s the name Drummoyne was applied to the whole vicinity. It was adopted as the name of the municipality in 1890. William Wright's mansion has since disappeared under the developer's hammer to make way for a block of white, brick home units. The remains of Wright's steps are marked with a council plaque.

Dry Creek *SA* is an industrial suburb and a creek in the northern metropolitan area of Adelaide. The name first appeared on Colonel Light's 1839 survey map and Light noted it was 'dry in summer'. Robert Milne, a young Scotsman, took up land on the upper reaches of Dry Creek in 1843. Between 1845 and 1850 he built his home, 'Drumminor', which still survives. Today it is used as a restaurant. Milne was one of the first farmers in the colony to use wire fences. Milne Road in the area is named after him. According to Rodney Cockburn, 'in 1916, the Yatala District Council discarded the name Dry Creek as applied to the old post office in favour of Pooraka, an Aboriginal word meaning "dry creek"'.

Dubbo *NSW* was the name given to a station owned by R.V. Dulhunty in the 1830s. In various Aboriginal dialects, the word *dubbo* means 'red earth', 'foggy', and 'head covering'. Dubbo was proclaimed a village in 1849, a municipality in 1872 and a city in 1966.

Dudley *NSW* was named after Dudley, a county borough and market centre of Worcestershire, England.

Duffy *ACT* is called after Sir Charles Gavan Duffy 1816-1903) a Victorian politician who worked for federation.

Duffys Forest *NSW* is a locality in Warringah Shire. Duffys Forest is named after P.J. Duffy, who was granted land in the area in 1857. He built a road to Cowan Creek to transport timber and also erected a stone wharf.

Dulwich Hill *NSW*, a suburb of Marrickville, was formerly known as Wardell's Bush and Wardell Hill. It was named after the London suburb of Dulwich.

Dumaresq River *NSW, Q* The river was discovered by Allan Cunningham in 1827, who named it 'in honour of the family with which His Excellency, the Governor, is so intimately connected." Henry Dumaresq was Governor Darling's brother-in-law. The river, and subsequently the township, were probably named after Lady Darling.

Dundas *NSW*, a suburb in the City of Parramatta, was the name used in 1799 when a few grants at the northern end of the locality were described as being in the Dundas district. An earlier grant was made in 1794 to Reverend James Bain, chaplain of the New South Wales Corps. The Dundas Council Chambers was built on it. The small settlers in the area later sold their lands to men with more capital, including Gregory Blaxland and William Cox.

Dunedoo *NSW* is an Aboriginal word meaning 'swan'.

Dungog *NSW* is an Aboriginal word meaning 'clear' or 'bare hill'. It was also the name of a local tribe.

Dunk Island *Q* was named by Captain Cook in 1770 after his patron, George Montagu Dunk, Earl of Halifax.

Dunolly *V* is a township twenty-three kilometres north of Maryborough. In 1846, Arch McDougall named his run after his family home, Dunolly Castle in Scotland.

Duntroon *ACT* is a military college, which takes its name from the property of Robert Campbell, who received a grant there in 1834. One account says he was related to the Campbells of Duntroon Castle in Argyllshire. Another story relates that the name probably originated with the sailing vessel *Duntroon*, a picture of which is to be seen in the Commandant's room in Duntroon House.

Durack River *WA* Named after the Durack family who pioneered the cattle country of western Queensland, and later opened up grazing land in the Kimberley district.

Dural *NSW*, a locality in Hornsby Shire on the North Shore of Sydney, appears in early records as Douro, Dooral and even Dure Hill.

Duyfken Point *Q*, near Waipa, was named by Mathew Flinders in honour of the voyage of the *Duyfken*. Willem Jansz, the Dutch captain of the *Duyfken* (meaning 'Little Dove') made the first recorded European sighting of Australia at the Pennefather River, forty-five kilometres north of Weipa.

Dynnyrne *T* follows from a house built by R.L. Murray in the late 1820s, and its subsequent application to his whole estate. It was a self-styled corruption by Murray of Dunnerne (or Dunearn) in Fife, the seat of the first baronet, Sir William Murray, his claimed ancestor.

Eacham, Lake *see* **Lake Eacham**

Eagle Farm *Q* is a locality in the Brisbane metropolitan area and gets its name because of the large number of eagles in the area. It was the site of a convict farm for women in the early days of settlement.

Eaglemont *V* is a suburb of the City of Heidelberg in the Melbourne metropolitan area. It was formerly Mount Eagle which was the northern boundary of a station taken up by Thomas Walker in 1838 and named 'The Eyries'.

Earlwood *NSW* is a suburb of the Sydney municipality of Canterbury. It is one of the most recent centres to develop in the municipality. Known early as Parkestown, then as Forest Hill, it became Earlwood about 1905. It is said to have been named after Earl, one-time mayor of Bexley who lived on the Bexley side of Wolli Creek, and the Wood brothers, William and James, who had a pig and poultry farm in the locality. However it is more likely to have been derived from Mrs Jane Earl, who owned land there from 1883. It was subdivided in 1905 as the Earlwood Estate.

East Hills *NSW,* a locality in Bankstown in Sydney's western suburbs, puzzles some local residents as it is not east of Sydney and the land is not all that hilly. East Hills was first mentioned in the *Sydney Gazette* of 1 December, 1810. The settlement farm of Robert Gardiner, who was a tenant of Mr G. Johnston, was called by this name. There are several places in England called East Hills.

Eastern Creek *NSW* is a locality in the City of Blacktown on the western outskirts of Sydney. It is the eastern branch of South Creek. Where the creek crossed the Western Highway, the village which grew up became known as Eastern Creek.

Eastlakes *NSW* is a suburb of the Sydney municipality of Botany. It was known earlier in Sydney's history by the less attractive title of Botany Swamps. They were Sydney's third source of water and later supplied water for the Sydney Hydraulic Company which ran lifts and wool presses in the city from the 1880s to the 1950s. The lakes have been partly reclaimed for airport extensions and a golf course.

Eastwood *NSW,* a locality in the Municipality of Ryde, was probably named after a village in central England. William Rutledge first bought the property and called it 'Eastwood'.

Ebor *NSW* After Eboracum, a Roman town in England which eventually became the city of York.

Echo, Lake *see* **Lake Echo**

Echuca *V* is an Aboriginal word meaning 'meeting of the waters', an appropriate name for a city located at the junction of the Murray, Campaspe and Goulburn rivers. It was originally known as Hopwood's Ferry after Henry Hopwood, who founded the town in 1853. In the 1850s, Echuca was the busiest inland port in Australia and was sometimes known as New Chicago.

Echunga *SA* is an Aboriginal word for 'near', 'close by', or 'at a short distance'. Another possible explanation is that it comes from an Aboriginal word *eechunga* which is said to sound like the call of the rufous whistler which makes a noise like 'eechung' with the accent on the 'chung'. However the derivation 'near' is more likely. The township was laid out by Jacob Hagen, a wealthy Quaker who arrived in December 1839. He acquired the land from J. Barton Hack. Hagen also built the local hotel, the Hagen Arms. The earliest recorded sale of an allotment was in 1849 so it can be assumed that Echunga Village was already in existence before this date.

Eden *NSW* Named after the family name of Baron Auckland, Secretary for the Colonies at the time the town was planned in 1842.

Eden Hill *WA* is named after a farm which once existed in the area. It was one of the estates of Henry Brockman's 1892 subdivision.

Edenhope *V*, a township on the shore of Lake Wallace thirty kilometres from the South Australian border on the Wimmera Highway, has a name derived from the Hope brothers from the River Eden in Scotland.

Edensor Park *NSW* is a suburb of Fairfield on the south-western outskirts of the Sydney metropolitan area. It has only one historical reference to it and this is by J.H. Wilson who mentioned that 'Edensor" was an estate in the south-east of England.

Edgecliff *NSW*, a Sydney suburb, is named after the rocky cliff that has largely been quarried-out today.

Edgeworth *NSW* in the Newcastle area was originally a mining settlement called Young Wallsend. It changed its name officially in 1961 to honour Sir Edgeworth David (1858–1934), one of Australia's greatest geologists who discovered coal in the region in 1886. He was also a Polar explorer who accompanied Sir Ernest Shackleton's expedition to Antarctica in 1907-8.

Edithvale *V* is a suburb of the City of Chelsea in the outskirts of Melbourne. The name was adopted, from Edithvale Road, as the name of the district in 1921. It came from 'Edith Vale Farm' which was named by the farmer after his wife.

Edmondson Park *NSW*, a suburb of Liverpool, is named after John Hurst Edmondson (1914-1941), a local lad who became the first soldier in the Australian army to win a Victoria Cross during World War II.

Eidsvold *Q* was first discovered and settled in 1848 by Thomas Archer, a friend of the explorer Ludwig Leichhardt. Archer was born in Scotland but was Norwegian by adoption and he named Eidsvold after a town in Norway. Gold was first discovered in Eidsvold in 1858 by a shepherd known as Lodder Bill.

Eildon *V*, a township by Lake Eildon, is named after Eildon Hills in Scotland, a reputed burial place of King Arthur.

Eildon, Lake *see* **Lake Eildon**

Ekibin *Q* is a locality in the Brisbane metropolitan area and gets its name from a derivation of an Aboriginal word, *yeekeben*, which was given to the part of the creek where the Aboriginals obtained edible aquatic roots.

Elanora Heights *NSW* is a locality in Warringah Shire. Elanora Heights is said to come from an Aboriginal word which means 'home' or 'camp by the sea'.

Eleenbah *NSW* is an Aboriginal word for 'beauty'.

Elermore Vale *NSW* is a Newcastle suburb named after the Elermore Colliery. The name was gazetted in 1975.

Eliza, Mount *see* **Mount Eliza**

Elizabeth *SA* is a city about thirty kilometres north-east of Adelaide. In 1955, the Premier of South Australia, Sir Thomas Playford, announced that the new town planned by the South Australian Housing Trust would be called Elizabeth after Queen Elizabeth II.

Elizabeth Bay *NSW* is a locality in the City of Sydney. It is named in honour of Governor Macquarie's wife whose maiden name was Elizabeth Henrietta Campbell. Elizabeth Street in the city and Mrs Macquarie's Chair and Point are also named after her.

Ellengrove *Q* is a locality in the Brisbane metropolitan area and was named by the Queensland Place Names Board in 1952, after a relative of the subdivider.

Elliott, Port *see* **Port Elliott**

Elsternwick *V* is a suburb of the City of Caulfield in the Melbourne metropolitan area. Elsternwick comes from *Elster*, a German word for 'magpie', which was the name chosen for his house by Charles Hotson Ebden, a pioneer overlander of cattle and later treasurer of the colony.

Eltham *V*, a shire on the fringes of the Melbourne metropolitan area, is known as the home of artists, potters, writers and other creative people. Eltham takes its name from Eltham in England. It was originally named Little Eltham when Josiah Morris Holloway, the first private developer in the district, subdivided the land into building blocks in 1850. Eltham became a district in 1856. It was proclaimed a shire in 1871.

Elwood *V*, is a locality in the City of St Kilda in the Melbourne metropolitan area. Elwood is probably named after the English poet and historian Thomas Elwood, who was a friend of the poet Milton.

Embleton *WA* honours an early settler, George Embleton, who emigrated to Western Australia with his wife and three-year-old son, arriving aboard the *Atwick* on 22 October 1829. He came as a servant of Dr John Watley who settled in the Bayswater district. Embleton bought land in the town of Perth in 1835 on which he built the Commercial Hotel. The suburb of Embleton was developed by the State Housing Commission after World War II.

Emerald *Q* is a town and railway junction about 270 kilometres west of Rockhampton. The first local settler, P.F. Macdonald, established 'Emerald Downs' station in about 1860 and the town takes its name from the station.

Emerald *V* was originally called Main Range but later assumed the name of Emerald Creek, which was probably named after Jack Emerald, an early prospector, who was murdered there.

Emerton *NSW* is a locality in the City of Blacktown on the western outskirts of Sydney. It is named after William Frederick Emert, a native of Siglingen, Württemberg, Germany, who arrived in 1853. He commenced business as a storekeeper and postmaster at Mount Druitt in 1861.

Emu Bottom *V*, the oldest homestead in Victoria, was built by George Evans in 1836. The word 'bottom' is an old English term for a low-lying swamp.

Emu Plains *NSW*, a locality within the boundaries of the City of Penrith, was named as early as 1808 because there were so many emus there. The name originated with Captain Tench but in the form of Emu Island. In 1814 Governor Macquarie referred to 'Emu Plains (hitherto erroneously called Emu Island)'.

Encounter Bay *SA* is situated near the mouth of the River Murray south of Lake Alexandrina into which the Murray River flows. The bay takes its name from the meeting in 1802 between Matthew Flinders in the *Investigator* and the French explorer, Nicholas Baudin, in the *Le Geographe*. It was named by Flinders. Baudin gave it the name Mollien Bay. The Aboriginal name for the bay was Wirramula.

Endeavour Hills *V* is named after Captain Cook's ship the *Endeavour*.

Endeavour River *Q* is named after Captain Cook's ship, the

Endeavour. When the *Endeavour* struck a reef on 11 June 1770, Cook beached it at a river mouth for repairs and gave the river its name. The spot is now the site of Cooktown.

Enfield *NSW*, a suburb of Burwood and Strathfield, was granted to William Faithful who had arrived in Australia in 1792 as a private in the New South Wales Corps. The land was later owned by the former convict, Simeon Lord, who became one of Sydney's wealthiest merchants. In 1822, it was bought by W.H. Moore, who cleared much of the heavily timbered area for farming. The name Enfield clearly came from the Middlesex market near London but we do not know when or why it was adopted. Its earliest known use was in 1853 when the first Enfield Post Office opened in Richard Fulljames's store near St Thomas's Church.

Enfield *SA* is a suburb in the corporate city of Enfield, six kilometres north of Adelaide. It is named after Enfield in Middlesex, England, the birthplace of Mr Hecox, a vendor of cordials and part-proprietor of Poor Man's Section as the area was originally known. This name in turn was derived from the fact that a small band of men, instead of applying for government relief during a depression, took up land to develop. The area was thickly covered with native pines.

Engadine *NSW* in Sutherland Shire was the name chosen by Charles McAlister for his estate because the scenery reminded him of Engadine in Switzerland. When the railway station was opened in 1924 Mrs McAlister, then a widow, named it after her estate.

Enmore *NSW*, a suburb of Marrickville, was named after a house built by Captain Sylvester Browne in 1835. It was situated in large grounds opposite the Enmore Post Office. It was demolished in the 1880s.

Enoggera *Q* is a locality in the Brisbane metropolitan area and is an Aboriginal word meaning 'plenty of wind'.

Eppalock, Lake *see* **Lake Eppalock**

Epping *NSW*, a suburb of the Shire of Hornsby and the City of Parramatta, is built on lands granted to Captain William Kent in 1796 and his nephew, Lieutenant William Kent, in 1803 on the north-western portion of the Field of Mars Common. Captain Kent, together with Captain Waterhouse, purchased the first merino sheep imported from the Cape of Good Hope in 1796. When the railway through the district opened in 1886 a platform named Field of Mars was built. This was soon afterwards changed to Carlingford. When the first post office opened it was called East Carlingford. This caused confusion and in 1899 the name

Epping was chosen on the recommendation of William Midson, a well-known resident whose father was born in the village of that name in Essex.

Epping *V* is named after Epping Forest in England.

Erindale *SA* is a suburb within the corporate city of Burnside. Erin is of course a popular name for Ireland. Erindale was part of the estate of Mrs S.A. Cowan which was subdivided in 1912 and 1917.

Ernaballa *SA*, according to A.W. Reed, is an Aboriginal word meaning 'creek with water holes'. The Ernaballa Creek and Gorge were discovered by Ernest Giles in 1873. The creek he named Ferdinand after his patron, Ferdinand von Mueller, but during the anti-German hysteria of World War I, the Aboriginal name was reinstated. The Ernaballa mission was established near the creek in 1937. The Geographical Names Board of South Australia has no record of Reed's meaning. According to the *Adelaide News* of 8 April 1937, a Mr John Carruthers, who did extensive survey work in the Musgrave Ranges in 1888, said that he 'named the water hole after an Aboriginal, Ernaballa, who claimed that the country belonged to him. Ernaballa became one of the most useful natives during the eight months the survey depot was located there'.

Erskineville *NSW*, a Sydney suburb, was named after the Reverend George Erskine, an early Wesleyan minister who called his residence 'Erskine Villa'.

Escape Cliffs *NT* So named because of the narrow escape of two officers of the *Beagle* in 1829 when they were pursued by Aboriginals. When South Australia annexed the Northern Territory in 1863, preparations were made to establish a settlement on the northern coast. In August 1864, Colonel B.T. Finniss led the hopeful settlers in a fleet of three vessels and chose a site as Escape Cliffs in Adam Bay. But dissensions occurred, and the settlement was a failure. The Aboriginal name of the cliffs was Pater-purrer.

Esperance *WA* recalls the vessel *Esperance* (Hope) commanded by Huon de Kermadec, an officer of Admiral d'Entrecasteaux's squadron, when the French survey team put in there in 1792. D'Entrecasteaux named the bay after Kermadec's ship. The town-site of Esperance, gazetted in 1893, was named after the bay on which it is situated.

Essendon *V* is a mainly residential city about ten kilometres from the centre of Melbourne. Essendon may be named after Essendon King, an early resident of the district, who was the grandson of Governor King of New South Wales.

Other sources indicate it was named after the village in Hertfordshire in England. Local government began in the area with the creation of a borough in 1861. Essendon was proclaimed a town in 1890, and became a city in 1909.

Ettalong *NSW* is probably an Aboriginal name. It was formerly known as Bar Swamp and then Gittins Lagoon.

Ettamogah *NSW* An Aboriginal word meaning 'let us have a drink'. This was the name of a local vineyard.

Eucla *WA* was first recorded by Lieutenant B. Douglas RNR, President of the Marine Board of S.A. who surveyed a port here in 1867. Most authorities agree that the name is an attempt to put down in English the Aboriginal name for the bluff which phonetically would be written Yerclia, a corruption of the Aboriginal word *yer*, 'bright' and *caloya*, 'fire'. This description was applied to the place by the Aboriginals because the planet Venus rises bright and clear over the high sand-dunes where the fresh water is located. The Aboriginal name of the town-site of Eucla is Chiniala.

Eumundi *Q*, a town in Maroochy Shire, is named after the chief of the Kabi Tribe, which helped an expedition to rescue survivors of the ship *Stirling Castle* which was wrecked off Fraser Island in 1836.

Eureka *V*, at Ballarat, was the site of a rebellion in December 1854 by miners, led by Peter Lalor, indignant at what they believed to be unreasonable licence fees.

Euroa *V*, a township forty-five kilometres south-west of Benalla, is said to come from an Aboriginal word, *yerao* meaning 'joyful'.

Evandale *SA* is the name of two places in the state. 1. Evandale is a suburb of the corporate town of St Peters. It was established and auctioned in1881 but there is no record of the origin of the name in the Lands Department archives.
2. Evandale near Angaston in the Barossa Valley was named around 1850 by Henry Evans, a local landowner who called it after himself. He was married to Sarah Lindsay Angas, the second daughter of George Fife Angas.

Evatt *ACT* is named after Herbert Vere Evatt (1894-1965), a Labor deputy prime minister from 1946 to 1949, and president of the United Nations General Assembly from 1948 to 1949.

Eveleigh *NSW*, a Sydney suburb, was called after an estate owned by Lieutenant William Holden who named it after his birthplace.

Everard Park *SA* is a suburb within the corporate city of Unley which is located just south of central Adelaide. Everard Park perpetuates the name of the Everard family which pioneered the area. Dr Charles George Everard arrived in South Australia in 1836 and later became a member of the Legislative Council. He farmed his land in Unley from the time he received his grant in 1839 until his death in 1876. The land remained in the family until 1911 when part of the old farm was subdivided and the area was named Black Forest. In 1917 a further portion was subdivided and the area was named Forest Gardens. In 1921 a further subdivision was made and the area was called Everard Park.

Evindale *SA* is a suburb of the City of Burnside on the south-eastern outskirts of the Adelaide metropolitan area. Evindale borrowed its name from the Cowan Estate at Burnside.

Ewey Bay *see* **Yowie Bay**

Ewie Bay *see* **Yowie Bay**

Exmouth Gulf *WA* 'The inlet was named Exmouth Gulf, in compliment to the noble and gallant Viscount', wrote P.P. King in February 1818, while conducting a survey of the coast in the *Mermaid*. King had served under Exmouth in the Royal Navy.

Eyre, Lake *see* **Lake Eyre**

Eyre Peninsula *SA* is between the Great Australian Bight and Spencer Gulf. It is named after Edward John Eyre, who explored the coastal area of the peninsula in 1839-1840.

Fadden *ACT* honours Sir Arthur Fadden (1895-1973), a Country Party politician who served as Prime Minister of Australia from August to October in 1941 and as Federal Treasurer from 1949 to 1958.

Fairfield *NSW* is a city on the south-western outskirts of the Sydney metropolitan area. Fairfield derived its name from a place of the same name in England. Vance George, the honorary archivist of Fairfield Council, who has undertaken extensive research into the history of the names of the municipality, says there are several references on the origin of Fairfield's name. Unfortunately, they generally tend to differ, but practically all agree on the one aspect that the name was taken from the Fairfield Estate in Somerset, with which a trustee of the Mark Lodge Estate had had previous association. The Mark Lodge Estate was the home of Captain John Horsley and the homestead was situated on the site where Fairfield District Hospital now stands. The property was put in the hands of trustees who were responsible for its disposal. From 1841, the name Fairfield appeared regularly in newspapers and documents. In a railway survey plan of the 1850s, Mr Thomas Ware Smart's home is clearly shown as 'Fairfield House' and because of its close proximity to the proposed railway station, the homestead's name was given to the station when it was built in 1856. Derivations from Fairfield are Fairfield Heights, Fairfield West and East Fairfield.

Fairfield *V* is a suburb of the City of Northcote in the Melbourne metropolitan area. Its name comes from a town in Derbyshire England by C.H. James, the owner and subdivider of Fairfield Estate.

Fairlight *NSW*, a suburb of Manly, is named after 'Fairlight House' built by the 'father of Manly', Henry Gilbert Smith (1902-1886). Smith bought land in 1853 and built the original stone 'Fairlight House' which was demolished in 1939 and is now remembered by the name of the suburb Fairlight, as well as Fairlight Crescent and Fairlight Street.

Fairy Bower *NSW*, a locality in Manly, was so named for the fairy-like beauty of the orginal bushland near the little beach.

Fairy Meadow *NSW* takes its name from Fairy Creek which flows through the area. It was originally called Cramsville in 1887 and became Balgownie in 1909 after John Buckland's estate which was named after a place in Scotland. The railway station was renamed Fairy Meadow in 1956.

Fairy, Port *see* **Port Fairy**

Farm Cove *NSW* An experimental farm was laid out in this part of what are now the Botanical Gardens in Sydney. Plants and seeds obtained at Rio de Janeiro and the Cape of Good Hope were planted, but with poor results, probably because the first plantings, by the Governor's house, were made in February. There was only one farmer in the First Fleet, and the sandstone soil in the bay on the eastern side of Sydney Cove, near the Governor's house, was not suitable for the purpose. It was only when cultivation was commenced at Parramatta that satisfactory results were obtained.

Farrer *ACT* is named after William James Farrer (1845-1906), a pioneer in scientific wheat-breeding who developed disease-resistant wheats.

Fassifern *NSW* is named after a Scottish village near Loch Eil.

Faulconbridge *NSW*, a locality in the Blue Mountains west of Sydney, was named late in the nineteenth century by Sir Henry Parkes after his mother Martha, whose maiden name was Faulconbridge. At the time of the name being applied, Sir Henry was Premier of New South Wales.

Fawkner *V* is a locality of the City of Broadmeadows in the Melbourne metropolitan area. Fawkner takes its name from the district's most famous pioneer settler, John Pascoe Fawkner, the co-founder of Melbourne about 1839. Pascoe Vale was the name of John Pascoe Fawkner's farm.

Felixstow *SA* is a suburb of Adelaide. It takes its name from the Reverend Thomas Quintin Stow of the Colonial Missionary Society, who settled in the district in 1843, and built his home 'Felix Stow' there. Stow Church in Adelaide perpetuates his memory. The Adelaide suburb was named by Stow after Felixstowe, a village in Suffolk. The name means 'happy place'.

Fern Tree *T* The earliest record found of this name is a reference to the first licensee of the 'Fern Tree Inn' (Alfred Hall, 29 August 1861). The origin is obviously the profusion of the tree ferns which still exist in the area.

Fernberg *Q* is the name of Government House in Brisbane.

Ferntree Gully *V* is a suburb of the City of Knox on the outskirts of the Melbourne metropolitan area. Settlement began in the 1840s. Its descriptive name comes from the Ferntree Gully National Park's famous ferntree gully.

Ferryden Park *SA* is a suburb in the corporate city of Enfield in the Adelaide metropolitan area. It was a private subdivision and there is no official record of the origin of the name but it was probably called after Ferryden in Angus, Scotland.

Fiddletown *NSW* is a locality in the Shire of Hornsby on the outskirts of Sydney. It was named after the Henstock brothers and a man named Small who took up selections in the area. Each had a violin or fiddle. The name was gazetted in 1976.

Field National Park, Mount *see* **Mount Field National Park**

Findon *SA* is the name of two places in the state that share their name with the town of Findon in Sussex. 1. Findon is a suburb of the corporate city of Woodville in the Adelaide metropolitan area. The land was originally granted in 1839 to G. Cortis, who was from Worthing in Sussex in England. He subdivided his land to form the village of Findon sometime between 1839 and 1850. The City of Woodville's abridged history says it was laid out in 1848. The first lot was sold in 1850.

2. Findon is a town near Port Elliot in the Goolwa district. It was laid out by G.L. Liptrott sometime between 1854 when he first bought it and 1857 when the first town lot was sold.

Fingal Bay *NSW* was originally called False Bay. The entrance was often, in days of sail and steam, mistaken for the entrance into Port Stephens.

Finke River *NT* was discovered by John McDouall Stuart in 1860. He named it after William Finke, who helped to finance his exploration.

Fisher *ACT* honours Andrew Fisher (1862-1928) who was Prime Minister of Australia from 1908 to 1909, 1910 to 1913 and 1914 to 1915.

Fishermens Bend *V* is a locality in Port Melbourne. It got its name because fishermen used to live there in the 1800s because fish were plentiful in the bay nearby at that time.

Fishers Ghost Creek *NSW* flows through Campbelltown. In 1826, a Campbelltown resident named Fredeick George Fisher disappeared in puzzling circumstances. About four months later, a man named Farley staggered ashen-faced into the Plough Inn and said he had seen Fisher's ghost sitting on the sliprails of a fence pointing to a creek in a nearby paddock. Police investigated, and found Fisher's body buried in a shallow grave at the spot the ghost had indicated. A neighbour of Fisher's named George Worrall was arrested, tried and convicted of the murder. Before his execution, Worrall admitted his guilt. Since then there have been many reported 'sightings' of the restless ghost of Frederick George Fisher. According to William A. Bayley's official history of Campbelltown, the problem with the story is that although the court records corroborate the general outline of the events, they contain no reference to the ghost. The story about the ghost appears in several books and in several languages. Several different individuals are credited with viewing the ghostly apparition, the most popular being a farmer named John Farley. According to some versions, Farley was drunk at the time. Others claim he was known for his sobriety. Whatever the truth of the matter, the story of Fisher's Ghost has become part of Australia's folklore.

Fitzroy *V* is a city in the inner metropolitan area of Melbourne. With Collingwood, it was known as Newtown during the 1840s. Fitzroy was named in 1850 after Sir Charles Augustus

Fitzroy, Governor of New South Wales, 1846-55. Fitzroy became a city in 1878.

Fitzroy River *Q* was named by the Archer Brothers in 1853 after the then governor of New South Wales.

Five Dock *NSW*, is a suburb of the Sydney municipality of Drummoyne. It is named after Five Dock farm, a grant made to Surgeon John Harris by Governor King in 1806. Nobody is sure of the origin of this unusual name. One theory is that there were five natural crevices on this headland which gave the name Five Dock. Only two of these crevices still remain.

Flemington *NSW*, a suburb of Strathfield, was called after John Fleming, an original grantee.

Flemington *V* is the place where Australia's most famous horse race, the Melbourne Cup, is held each year. Flemington may have been named after Robert Fleming who supplied meat for race meetings there, but it is more likely that it is called after the Flemington Estate in England.

Flinders *V*, a coastal town on Mornington Peninsula, was named after the famous navigator, Matthew Flinders.

Flinders Island *T* was named by Governor King after Matthew Flinders, who circumnavigated Van Diemens Land in 1798.

Flinders Park *SA* is a suburb within the Corporation of Woodville, a city in the Adelaide metropolitan area. Flinders Park is one of the many memorials to Captain Matthew Flinders, and was subdivided first on 9 November 1923. The portion formerly known as Findon South and Underdale still exists. It is located on the south of Flinders Park in the Corporation of the City of West Torrens.

Flinders Peak *V* is a prominent peak on the You Yangs, a small range of mountains between Werribee and Geelong. Matthew Flinders climbed and named it Station Peak in May 1802. It was later renamed to honour Flinders.

Flinders Ranges *SA* are named after Matthew Flinders, the British explorer. Flinders sighted the peaks from Spencer Gulf in 1802, but it was 1839 before the explorer Edward John Eyre first visited the area.

Flinders River *Q* was discovered in 1841 by Captain J.L. Stokes of the *Beagle*, during a survey cruise of the Gulf of Carpentaria. He named it after Matthew Flinders.

Floreat *WA* is derived from the Perth City Council motto, *floreat* meaning 'let it flourish'. The name was originally proposed by W.E. Bold, a town clerk of Perth for over 44 years. The land was bought by the Council in 1917 and was originally called Floreat Park. It was changed to Floreat in 1977.

Florey *ACT* honours Baron Howard Walter Florey (1899-1968), an Australian scientist who shared the Nobel Prize in 1945 for his work in penicillin research.

Fly Point *NSW* was once the site of Chines fish-curing facilities, customs buildings and the first school. During World War II it was established as an armed forces personnel base. Today, a sports oval, high school and hospital are situated here. The name is derived from a nautical term meaning 'safe anchorage with protection from winds'.

Flynn *ACT* is called after John Flynn (1880-1951), the Presbyterian minister and missionary who was mainly responsible for the formation of the Australian Inland Mission Aerial Medical Service in 1928, the forerunner of the Royal Flying Doctor Service of Australia.

Footscray *V* is a suburban and industrial city on the Maribyrnong River. It is located about six kilometres from the centre of Melbourne. Footscray takes its name from Footscray in Kent, England. Footscray became a town in 1889. It was declared a city in 1891.

Forbes *NSW* is named after Sir Francis Forbes (1784-1841), the first Chief Justice of New South Wales. George William Evans explored the region in 1815 and John Oxley visited it in 1817.

Forrest *ACT* honours two West Australian brothers, Sir John Forrest (1847-1918), an explorer who later became a premier of Western Australia and served in the first federal parliament in 1901, and Alexander Forrest (1849-1901), who led an expedition in 1879 that opened up valuable country in the Kimberley and Fitzroy districts of Western Australia.

Forrestfield *WA* was named by one of the early settlers, Charles Hale, who established a farm there in 1902, planting his crop under the trees. The story goes that a neighbour was amused that Hale was watering and cultivating a forest and a field at the same time. The name 'Forest Field' is said to have been coined by the neighbour for Hale's property. As the name is spelt with a double R it is more likely that the name has some connection with the Forrest family. Alexander Forrest once owned the land there.

Forster *NSW* was named by William Forster (1818-82), Premier of New South Wales from 1859 to 1860. Despite the way it is spelt, it is almost universally pronounced Foster in New South Wales, presumably because people with the surname, Forster, usually pronounce it Foster.

Forth River *T* was originally called 'Third Western River' by Van Diemen's Land Company Surveyor, Alexander Goldie, in 1826, but by 1830 the present name had been given, presumably by that Company, after the river in Scotland.

Fortitude Valley *Q* is a locality in the Brisbane metropolitan area and was named in 1849 after a ship that brought immigrants to Moreton Bay.

Forty Baskets Beach *NSW*, a locality in Manly, is said to have got its name because forty baskets of fish were once caught there.

Foster *V* a township north-west of Corner Inlet, was named in 1871 after W.H. Foster who was magistrate and warden of the South Gippsland goldfield. It was formerly known as Stockyard Creek.

Fowlers Bay; Fowler Point *SA* were named by Flinders in 1802 after the First Lieutenant on the *Investigator*, Robert Fowler, who later in life rose to the rank of admiral. The bay and point were seen by Peter Nuijts in 1627. A town-site was dedicated in 1890 and given the Aboriginal name 'Yalata' meaning *shellfish*, but it reverted to Fowlers Bay in 1940.

Franklin River *T* was named after Sir John Franklin, a Lieutenant-Governor of Tasmania, who was also known for his polar exploration and his discovery of the North-West Passage.

Frankston *V* is a city on the Mornington Peninsula on the eastern shore of Port Phillip Bay about forty kilometres south-east of Melbourne. There are several theories concerning the origin of the name of Frankston. The most popular version is that it was named after Frank Stone's Hotel, one of the earliest buildings in Frankston. Another theory suggests it was called after one of the first settlers, Frank Liardet, who took up land there in 1843. His house, 'Ballam Park', built in 1850, still stands in Cranbourne Road. Frankston was created a district in 1860 and became a shire in 1871.

Fraser *ACT* is named in honour of James Reay Fraser (1908-1970) who was Member of Parliament representing the Australian Capital Territory from 1957 to 1971.

Fraser Island *Q* is the largest sand island in the world. In 1836, survivors from the wreck of the *Stirling Castle* came ashore there. The party included captain James Fraser, who was later killed by Aboriginals. The island was named after him.

Frederick Henry Bay *T,* which forms the eastern boundary of the Clarence Municipality, was named by Abel Tasman in 1642 after the then Stadtholder of the Dutch United Provinces.

Freeling *SA* was established in 1860 and named after Major-General Sir Arthur Henry Freeling, a surveyor-general and colonial engineer who succeeded Colonel Frome in 1849. Freeling was also a member of the first elected Legislative Council. He retired as Surveyor-General in 1861 and returned to England.

Fremantle *WA* is the chief port of Western Australia, and serves as a harbour for the city of Perth which is located about twenty kilometres away. Fremantle is named after Captain Charles Howe Fremantle who, on 2 May 1829, planted the Union Jack on Arthurs Head and took formal possession, in the name of His Majesty King George IV of 'all that part of New Holland which is not included within the territory of New South Wales.' On June 1, 1829, Captain James Stirling arrived in the *Parmelia* with a party of sixty-eight new settlers to found the Swan River Colony. He chose the name, Fremantle, to honour Captain Fremantle.

French Island *V* in Western Port was named Ille de Francais by the French Captain Baudin when he camped there in 1802 while on his scientific expedition around the coast of Australia.

Frenchs Forest *NSW* is a locality in Warringah Shire. Frenchs Forest bears the name of James Harris French, who acquired land in the area in 1856 and developed a timber industry with two sawmills. Large-sized trees were cut and split by hand and then hauled by bullock to a wharf at Bantry Bay.

Freshwater *NSW* is a locality within the boundaries of Manly municipality. The first grant of land in the area was made to Thomas Brain in 1818. Evidence suggests that Brain's former land took the name Freshwater Estate about 1884. Local opinion is divided as to whether the name originated from Freshwater on the Isle of Wight, England or from the small freshwater stream that rose above Oliver Street and flowed into the sea at the northern end of the beach. Freshwater and Harbord began as adjoining localities at the time Lord Carrington became Governor of New South Wales in 1885. However, the growing popularity of surf bathing early in the twentieth century gave beachside Freshwater prominence over its western neighbour,

Harbord. In 1923 the whole area was named Harbord (the maiden surname of Lady Carrington). In 1980 the Geographical Names Board re-named Harbord Beach and its adjoining reserve, Freshwater.

Freycinet Peninsula *T* was named by Captain Nicolas Baudin after his cartographer, Henri Freycinet, while charting the coastline in 1802.

Frome, Lake *see* **Lake Frome**

Fulham *SA* is a suburb located within the Corporation of the City of West Torrens in the Adelaide metropolitan area. It was first settled in 1836 by John White who called his farm Fulham Farm because of his association with Fulham on the banks of the Thames in London. Fulham was first subdivided in 1877.

Fulham Gardens *SA* is a suburb within the boundaries of the corporate city of Woodville in the Adelaide metropolitan area. The area was once part of John White's Fulham Farm. The name was approved by the Nomenclature Committee in 1925.

Fyansford *V* is a rural and industrial locality five kilometres west of Geelong near the junction of the Moorabool and Barwon rivers. Captain Foster Fyans, Crown Lands Commissioner and Police Magistrate at Geelong, had his house 'Balyang' on the Barwon, downriver from a ford once used by western travellers.

Fyshwick *ACT* is named in honour of Sir Philip Fysh, one of the founders of the constitution.

Gabo Island *V* was named by Captain Cook on 20 April 1770 and is south-west of Cape Howe, near the border of Victoria and New South Wales. There is a story, probably apocryphal, that when asked the name of the locality, the Aboriginals replied *gabo*, 'we don't understand'.

Gagebrook *T* is called after a house built by John Frederick Gage in 1880 after an earlier building was destroyed by fire. His father, John Ogle Gage, took up his grant of land in 1824.

Gambier, Mount *see* **Mount Gambier**

Garden Island *NSW*, in Sydney Harbour, got its name because it was cleared on 11 February 1788 for a vegetable garden for HMS *Sirius* and HMS *Supply*. It was reserved for naval use in 1800. A new dockyard was built in 1939. It suffered a Japanese attack on 31 May 1942 when the ferry *Kuttabul* was torpedoed.

Garden Vale *V* is a suburb of the City of Caulfield in the Melbourne metropolitan area. Garden Vale is of unknown origin but may reflect the fact that there were once many market gardens in the area.

Gardiner *V* is a suburb of the City of Malvern in the Melbourne metropolitan area. Gardiner was the original name of the area settled by John Gardiner who travelled overland from New South Wales in 1835 with about 400 head of cattle and set up a cattle station.

Garie *NSW*, in Sutherland Shire, may be an Aboriginal word meaning 'sleepy' or it may not be an Aboriginal word at all.

Garran *ACT* honours Sir Robert (Randolph) Garran (1867-1957), a federal politician who helped to draft the Australian constitution. He played an important role in the community of Canberra in its early days and helped to found Canberra University College.

Gascoyne River *WA* was discovered by George Grey in 1839 and named by him 'in compliment to my friend, Captain J. Gascoyne'.

Gatton *Q* is a town on Lockyer Creek, about ninety kilometres west of Brisbane. The site for the village of Gatton was gazetted in 1855. It was probably named after Gatton in Roxburghshire, Scotland.

Gawler *SA* was named after George Gawler who was Governor of South Australia from 1838 to 1841. The town was laid out in 1839 by William Jacob, from a plan prepared by Colonel William Light, who had visited the area in 1837. The local government at Gawler adopted the Governor's coat-of-arms as its insignia and crest. The Aboriginal name for the area is Kaleeya or Kaleteeya.

Gayndah *Q* Though undoubtedly of Aboriginal origin, there is some uncertainty about the name. It comes either from *gu-in-dah* or *gi-un-dah*, 'thunder', or from Ngainta, 'place of scrub'. It was called Norton's Camp at one time, after a local carrier.

Gaythorne *Q* is a locality in the Brisbane metropolitan area and was named by the Railways Department in 1923 after a nearby property.

Geebung *Q* is a locality in the Brisbane metropolitan area and is named after the fruit of a wild shrub.

Geelong *V* on Corio Bay is the largest provincial city in Victoria. It was incorporated in October 1849 and proclaimed a city on 8 December 1910. It was named in 1837 by Governor Bourke using an Aboriginal name for the district, *Jillong* (noted as early as 1924 by Hume and Hovell). Various meanings have been given. They include 'place of the cliff' (i.e. East and West Beaches of today) or 'white seabird', or 'swamps where native companions live'. The first or second meaning is usually accepted.

Geikie Gorge *WA* was named in 1883, in honour of a famous British geologist Sir Archibald Geikie, by Edward Hardman who was the first geologist to examine the Kimberley district.

Gellibrand *V* is named after Joseph Tice Gellibrand, who drew up the deed of purchase for the site of Melbourne at the time of Batman's purchase. The Aboriginal name was Walar Walar.

Gembrook *V*, a township in the Dandenong Ranges south-east of Melbourne, was named by the Revd Dr Bleesdale who found gemstones in the creek.

Geographe Bay *WA* 'We discovered a very large bay open to the north-west, which was given the name Geographe Bay', wrote Lieutenant de Freycinet who was on the *Naturaliste* at the time. It was on 30 May 1801 that Captain Nicolas Baudin anchored *Le Geographe* in the bay during his scientific expedition, and lost contact with the *Naturaliste*.

George, Lake *see* **Lake George**

George Town *T* was first known as Outer Cove and then as York Cove. It was given its name in honour of King George III in 1811 when it was declared a town.

Georges Hall *NSW*, near Bankstown, a city located in the western part of the Sydney metropolitan area, was named after King George III.

Georges Heights *NSW*, a locality in Mosman on Sydney's North Shore, was named after King George III.

Georges River *NSW* was named after King George III, probably by Governor Phillip.

Georgina River *Q* was discovered by William Landsborough in 1861, while travelling south from the Gulf country along Burke and Wills' route. Originally it was called the Herbert River in honour of the first premier of Queensland. Landsborough later renamed it Georgina after the daughter of Sir Arthur Kennedy, Governor of Queensland from 1877 to 1883.

Gepps Cross *SA* is a locality in the corporate city of Enfield in the Adelaide metropolitan area. The crossing place was a five-road junction where Isaac Gepp, an early land-owner, kept the Gepps Cross Hotel, in the early days of the colony.

Geraldton *WA* was named after Governor Charles FitzGerald, when the town-site was surveyed in 1850. Lead was discovered on the Murchison River by a party under A.C. Gregory in 1848, the mine that was established also being named for the Governor. This place was originally known and recorded as Champion Bay since the naming of the bay at the head of which it stands, by J.L. Stokes in 1840.

Gibson Desert *WA* was named in honour of Alfred Gibson, who died in 1876 during Ernest Giles' ill-fated second attempt to cross it. The first European crossing was made by John Forrest in 1874.

Gidgealpa *SA* was chosen by Jeff Green, an American resident manager with Delhi Australian Petroleum Ltd. in 1963. He took it from the name of a waterhole near the gas field, believing that it meant 'woman standing under a grey rain cloud.' In fact, the waterhole was named by members of the Jandrawanta tribe and was called Kilyalpa which means 'to stand in the shade of a grey rain cloud'. Kilyalpani was one of the mythical women who created the land. She once prayed for rain and while doing this a grey cloud appeared above her so that she stood in its shade.

Gilberton *SA* is a suburb within the corporate town of Walkerville in the inner metropolitan area of Adelaide. Gilberton is named after Joseph Gilbert who bought the land in 1846. He never lived in the area that perpetuates his name although he did live in Adelaide for a time. Gilberton was surveyed and sold in the early 1870s.

Gilgandra *NSW* is Aboriginal for 'long waterhole'.

Gilles Plains *SA* is a locality partly in the City of Tea Tree Gully and partly in the corporate city of Enfield in the Adelaide metropolitan area. Gilles Plains was named after Osmond Gilles, the first Colonial Treasurer of South Australia. 'Sudholz Farm' at Gilles Plains was one of the best known properties in the early decades of the colony, having been established by a Mr Martin. He sold to G.W.W. Sudholz who, like Ragless, Bowman and Folland, was a benefactor to many workless labourers in the depression periods of the 1800s. Their names are commemorated in the street names of the north-east region of the Adelaide metropolitan area.

Gilmore *ACT* is named after Dame Mary Gilmore (1865-1962), an Australian poet.

Gin Gin *Q* is a corruption of Jinjin burra, the local Aboriginal tribe.

Gippsland *V*, the south-east region of Victoria, was originally called Caledonia Australis by the explorer Angus McMillan, who visited the area in 1840 and 1841. It was later renamed Gippsland in honour of Governor Gipps by the Polish explorer, P.E. de Strzelecki.

Giralang *ACT* meaning 'star' comes from the language of the Wiradhuri Aboriginal tribe of the Central West of New South Wales.

Girraween *NSW*, a suburb of the Municipality of Holroyd in Sydney's western suburbs, is an Aboriginal word meaning 'place of flowers'.

Gisborne *V*, a township fifty-two kilometres north-west of Melbourne, was named by Governor La Trobe after Henry Fysche Gisborne who was a Crown Lands Commissioner.

Gladesville *NSW*, a locality in the Municipalities of Ryde and Hunters Hill, comes from the pioneer of the area, John Glade, a convict who arrived in Australia in 1791 to serve a seven-year sentence. He died in 1848 and was buried at St Anne's cemetery.

Gladstone *Q* is a city on Port Curtis, 529 kilometres, by road, north of Brisbane. The Aboriginal tribe that lived in the Gladstone area was the Biele or Byellel. Matthew Flinders visited Port Curtis in 1802 and named it after Admiral Sir Roger Curtis. One of Gladstone's early settlers, Joseph Willmott, named the river the Liffey. In 1854 Governor Sir Charles Fitzroy renamed it the Calliope, after the ship in which he arrived with his official party of sixty settlers to install Sir Maurice O'Connell as first government resident. The town was named after Sir William Gladstone, the famous British prime minister.

Glasshouse Mountains *Q* are steep, cone-shaped hills about eighty kilometres north of Brisbane. Geologists call them volcanic plugs. Captain Cook saw them in 1770 while sailing up the east coast of Australia. He called them the Glasshouse Mountains because their shape reminded him of the glass factories of his native Yorkshire.

Glebe *NSW* was set aside for church purposes by Governor Phillip in 1789. Glebe means an area of land devoted to the maintenance of an incumbent of the church.

Glen Innes *NSW* was named after Archibald Clunes Innes (1800-57), a soldier and pioneer, who was granted land when he was appointed police magistrate at Port Macquarie. By means of convict labour he transformed a wilderness into the best pastoral property north of Sydney, and named it Lake Innes. He owned many properties, including Furracabad, the site of Glen Innes, which was named by him, though he never lived there. The Aboriginal name was Kindaitchin, meaning 'plenty of stones'.

Glen Iris *V* is a locality between the Melbourne suburbs of Camberwell and Malvern. Glen Iris comes from a house built by a solicitor named J.C. Turner. The name for the area was adopted when Glen Iris Estate was sold in 1861.

Glen Osmond *SA* is a suburb of the City of Burnside on the south-eastern outskirts of the Adelaide metropolitan area. It takes its name from Osmond Gilles, the first Colonial Treasurer, 'who died there on 23 September 1866, aged seventy-nine', according to Rodney Cockburn in his book *What's In A Name*. Gilles bought the land and named it on 19 October 1839. The adjoining suburb of Mount Osmond has the same origin. The silver-lead mine at Glen Osmond was the first mineral discovery in the state.

Glen Waverley *V* is a suburb of the City of Waverley in the Melbourne metropolitan area. Glen Waverley, formerly known as Black Flat, was gazetted as a township in 1905.

Glenalta *SA* There is no record of origin. It could mean 'high valley'.

Glenbrook *NSW* is a locality in the Blue Mountains west of Sydney. It was formerly known as Wascoe's Siding. In 1874 a passenger platform was added and in 1878 the name was changed to Brookdale. In 1879 the name was altered to Glenbrook.

Glenelg *SA* is a city at Holdfast Bay on the eastern shores of Gulf St Vincent, ten kilometres south-west of the city of Adelaide. Governor Hindmarsh named Glenelg after Lord Glenelg (1788-1866), who was Secretary of State for the Colonies when the province of South Australia was founded. Glenelg is one of only two palindromes recorded in South Australia nomenclature. Navan is the other one. The Aboriginal name for Glenelg was Cowiandilla, which means 'the place of drinking water'. New Glenelg was known to the Aboriginals as Patawilya meaning 'swampy and bushy with fish'. The enrolled plan approved by Governor Gawler shows the Patawalonga (another form of Patawilya) as the River Thames. The Aboriginal name for the Old Gum Tree (where the Proclamation of the Province is said to have been read) was *pudtha yukoona*, meaning 'arched red gum'.

Glenelg River *V* flows from the Grampians to Discovery Bay in Western Victoria. It was discovered by T.L. Mitchell in 1836 and named in honour of Lord Glenelg, Secretary of State for the Colonies.

Glenfield *NSW* takes its name from a grant given to Dr Throsby. 'Glenfield House' still stands and is one of the historic homes of the City of Liverpool on the outskirts of present-day Sydney.

Glenhaven *NSW* is a locality in the Baulkham Hills Shire on the north-western outskirts of Sydney. Glenhaven was originally known as Sandhurst but the name caused confusion as there was a town of the same name in one of the other states. It was then decided to change the name. At a public meeting held in 1893 or 1894 the present name was chosen. The lower portion of the settlement was known as The Haven and as the village lay in a valley or glen, Glenhaven was considered appropriate.

Glenhuntly *V* is a suburb of the City of Caulfield in the Melbourne metropolitan area. Glenhuntly comes from a ship named *Glen Huntly* which arrived in Port Phillip in 1840.

Glenorchy *T* was visited by Governor Macquarie in 1811. He is believed to have named Glenorchy after a village on the Orchy River in Scotland. The modern spelling is said to have come from a corruption of the Gaelic word *gleann* meaning 'tumbling waters'.

Glenrowan *V*, a township sixteen kilometres south-west of Wangaratta, was named by the brothers Rowan who had a station there. In June 1880 the Kelly gang of bushrangers made their last stand at Glenrowan.

Glenroy *V* is a suburb of Broadmeadows City in the Melbourne metropolitan area. Glenroy was named, after Glenroy in Inverness, Scotland, by Duncan Cameron, one of the pioneers who brought sheep to the area between Moonee Ponds and Merri Creek.

Gloucester *NSW* takes its name from the Gloucester Estate which was owned by the Australian Agricultural Company in the late 1890s. The obvious origin is in the cathedral city of Gloucester in Gloucestershire, England.

Goat Island *NSW* in Sydney Harbour may have got its name from three goats which were brought with other livestock from Capetown in the First Fleet in 1788. Unfortunately, there is no documentary proof of this theory. The island was later used to house convicts. It is now owned and used by the Maritime Services Board.

Gold Coast *Q* is a city that stretches over forty kilometres along the coast, from Coolangatta on the New South Wales border to Paradise Point in the north. It attracts nearly three million visitors a year. Its best known area is Surfers Paradise. The name, Gold Coast, reflects the rapidly rising land values of the area. The town of the Gold Coast was established in1958, and the City of the Gold Coast was established in 1959.

Golden Grove *SA* is a suburb in the corporate city of Tea Tree Gully in the Adelaide metropolitan area. It derived its name from the property 'Golden Grove Farm' owned by Captain Adam Robertson, who bought land in the area from the South Australian Company in 1842. By 1853, he had built 'Golden Grove House'. Captain Robertson called his property after his ship *Golden Grove* and was incensed when the name was adopted for the district generally. He wrote to the Postmaster General in July 1859 protesting strongly against its application to the local post office and claimed it exclusively for his farm. However, a press correspondent pointed out that several years before he had conferred the name 'Golden Grove' upon the school, and therefore he could not blame the settlers for extending it to the village as well.

Good Island *Q* First named Good's Island by Flinders on 2 November 1802 after the 'botanical gardener' on the *Investigator*. The name was subsequently misspelt Goode, but in 1948 the Admiralty changed it back to Good. The Aboriginal name was Palilug.

Goodnight *Q* This peculiar name was derived from an incident when the captain of a river steamer on the Murray heard a voice calling from the bank, 'Goodnight!' Ever afterwards he called that particular spot 'the place where the bloke said "goodnight" '.

Goolwa *SA* comes from an Aboriginal word which may mean 'elbow'. It is also recorded that *goolwa* is Aboriginal for 'mixed water', sometimes fresh from the Murray River and sometimes salty from the sea. It was originally surveyed in 1839 and named 'town-on-the-Goolwa' by Captain E.C. Frome, Light's successor as Surveyor-General. It was formerly known as Port Pullen, after Captain Pullen, Colonel Light's second-in-command, who later became an Arctic explorer. In 1853, on the instructions of Governor Young, another town was surveyed in the adjoining land

121

to the south of the original township. The new town was called Goolwa. The survey of the new town was carried out in conjunction with the construction of a tramway from Port Elliott to Goolwa to enable cargo from the riverboats to be transferred to shipping at Port Elliott.

Goondiwindi *Q* comes from an Aboriginal word meaning 'droppings of ducks or shags', 'place of wild ducks', or 'water running over rocks'. It is pronounced Gun-der-windy. It was the home of the famous racehorse Gunsynd.

Goonoo Goonoo *NSW* comes from an Aboriginal word meaning 'green tree', or 'plenty of water', or 'poor game-land'. The way it is written bears no relationship to the pronunciation which is Gunner Gernoo.

Goose Island *WA* Flinders' men killed sixty-five Cape Barren geese here in January 1802.

Gooseberry Hill *WA* is in Kalamunda Shire on the outskirts of the Perth metropolitan area. It has been recognised as a locality since 1860 when Benjamin Robins, a farmer of Guildford, took up land there. The name appears to have its origin in the fact that cape gooseberries thrived in the area.

Gordon River *T* was discovered by James Kelly in 1815 during a voyage of exploration around the island in a whale boat lent to him by the Pittwater magistrate, James Gordon, after whom Kelly named the river.

Gordonvale *Q* was originally called Mulgrave and then Nelson. It was renamed Gordonvale in 1912 after John Gordon, a pioneer of the area.

Gore Cove *NSW*, a locality in the Municipality of Lane Cove, is a sheltered inlet near Greenwich. It is named after William Gore although his property was at Artarmon, some distance away. However, in addition to his Artarmon estate inland, William Gore had five hectares of land granted to him on the Greenwich side of Gore Cove which perhaps gave it its name.

Gore Hill *NSW* is the site of the ABC television tower on Sydney's North Shore. The name commemorates William Gore, one of the early settlers and a former Provost Marshall. After the New South Wales Corps arrested Bligh in the famous Rum Rebellion of 1808, they also arrested Gore. He was sentenced to transportation for several years and sent to the Coal River (Newcastle) where he worked alongside ordinary convicts. In 1810, Governor Macquarie restored Gore to his former office. He was given land on the North

Shore. His grant included an area bounded on modern maps roughly by Mowbray Road, Elizabeth Street to Artarmon Station and by a line north from Chelmsford Avenue back to Mowbray Road.

Gormanston *T* is named after Governor Jenico William Joseph Preston, 14th Viscount Gormanston.

Gosford *NSW* was first referred to as the township at Point Frederick. Then in February 1839, when the plan was sent to Governor Gipps for approval as the Township at Brisbane Water, the plan was returned by the Governor in April, marked as the Plan for Gosford with no explanation to indicate the reason for its name. Research into the background of Governor Gipps' life revealed that he had served as Commissioner to Canada with Sir George Grey and the Earl of Gosford from 1835 to 1837 just before being appointed Governor of New South Wales. It seems that Gipps took the opportunity to honour his friend by using his name for the township.

Gosnells *WA* is a city in the Perth metropolitan area. Gosnells gets its name from Charles Gosnell, a director of the London firm of John Gosnell and Co., perfume manufacturers, who bought a farm in the district in 1862.

Gosse Range *NT* was named after Mr Harry Gosse by Ernest Giles in 1872. William Christie Gosse came into the area in 1873 and named Ayers Rock and became Deputy Surveyor-General in 1875.

Goulburn *NSW* was originally known as Strathallen. Surveyor John Meehan, who visited the area in 1818, named the region in honour of Henry Goulburn, who was then Secretary for the Colonies.

Goulburn River *V* The Goulburn was crossed by Hamilton Hume and Henry Hovell in 1824 and named the Hovell, but was later renamed in honour of Frederick Goulburn. In 1836, T.L. Mitchell complained that the name 'Hovell' had not been retained as there was already a Goulburn River in New South Wales. The Aboriginal name was Bayunga.

Gove Peninsula *NT* was a base for 5000 servicemen during World War II. Gove Airport was named in honour of Pilot Officer William J. Gove, who was killed in action in 1943.

Govetts Leap *NSW*, a locality in the Blue Mountains west of Sydney, was named by Major Mitchell after his assistant, a surveyor named William Romaine Govett. A more colourful, but unfortunately inaccurate story, is that it was named after a convict named Govett, who jumped over it. This story is contained in a letter written by Governor-General Denison in 1858.

Gowerville *V* is a suburb of Preston, a city in the Melbourne metropolitan area. It was bought by an agent for Abel Gower of London, in land sales in 1838.

Gowrie *ACT* honours the Earl of Gowrie, Brigadier-General Alexander Gore Arkwright Hore-Ruthven (1872-1955), who was Governor-General of Australia from 1936 to 1944.

Gowrie Park *V* is a locality in the City of Broadmeadows in the Melbourne metropolitan area. Gowrie Park was the name of the estate owned by a man called Robinson who came from Gowrie in England.

Gracemere *Q* Named by the discoverers, Charles and William Archer, in May 1853 as a compliment to Mrs Grace Lindsay Morison. The lagoon was first called Farris.

Graceville *Q* is a locality in the Brisbane metropolitan area and means 'agreeable village'.

Grafton *NSW* was originally called The Settlement. Governor Fitzroy renamed it after his grandfather, the Duke of Grafton, who was Prime Minister of England from 1767 to 1770. It became a municipality in 1859 and a city in 1885.

Grampians *V* In July 1833 T.L. Mitchell came in sight of the mountains and named them after their Scottish counterpart. Aboriginal names for the range were 'Cowa' and 'Narram Narram'.

Grange *SA* is a suburb in the western area of the Adelaide metropolitan area. Its name comes from the house built by the explorer, Captain Charles Sturt, and from a London suburb. Captain Sturt lived in 'The Grange' with his family between 1840 and 1853. It was from this home that Captain Sturt set out on 15 September 1844 to test the theory of an inland sea in the heart of the continent. To this house he returned on 19 January 1846, exhausted by a journey during which he was almost given up for lost. Although compelled eventually to live in England, he would not sell 'The Grange', which remained his property until his death in 1869. This old house is situated just off Jetty Street, Grange, close to the Grange Primary School.

Granville *NSW*, a suburb of the City of Parramatta, is named after the Earl of Granville (1815-1891) who was Foreign Minister from 1870 to 1874 and again from 1880 to 1885. It was called Parramatta Junction until 1880 when the local citizens decided to change it.

Gravatt, Mount *see* **Mount Gravatt**

Graylands *WA* was probably named after David Gray from Middlesex, England who arrived in the colony on 14 October 1854. He was a carpenter, bricklayer and stonemason and built the Cemetery Road (Forrest Avenue), Perth. He owned a small farm near Claremont but it is also possible Graylands was named after Mrs Maria Gray, wife of a warden of Fremantle Prison. She lived in Fremantle and let the property she owned in the Graylands area. The name was first recorded in 1902 when Graylands Road was gazetted. The first house in Graylands was completed in March 1945. It heralded the beginning of a joint State-Commonwealth Government housing project.

Grays Point *NSW* in Sutherland Shire is the subject of two theories. The first, which is more likely, is that it was named after Samuel William Gray who owned land on the point. The second is that it was named after John Edward Gray who was a resident ranger in the National Park in the late 1800s. He lived at Gundamaian and became a well-known local identity.

Great Australian Bight *SA, WA* Although the first Europeans to sail along the coast were Pieter Nuijts in the *Gulden Zeepaard* in 1627, and later d'Entrecasteaux in 1792, the name was not given until 1802, when Matthew Flinders referred to it as the 'great bight or gulph of New Holland', and in his *A Voyage to Terra Australis* as the Great Australian Bight.

Great Victoria Desert *SA, WA* Crossing the desert from east to west in 1875, Ernest Giles travelled nearly 500 km without finding water until he reached a supply which he named the Great Victoria Spring. The desert was given the same name, doubtless after Queen Victoria. It is located north of the Nullarbor Plain.

Great Western *V* is a locality north-west of Ararat. There is uncertainty whether the name was bestowed on account of the geographical position or not. 'Great' probably expressed diggers' optimism for the goldfield's gold-bearing qualities.

Greenacre *NSW* is a suburb of Bankstown, a city located in the western part of the Sydney metropolitan area. Greenacre was an acre of clear land off Liverpool Road, near the present-day Stacey Street, developed by Arthur Rickard and Co. Ltd. in the early twentieth century.

Greenacres *SA* is a suburb in the corporate city of Enfield in the Adelaide metropolitan area. Greenacres is a descriptive name. This land was originally granted to Duncan Dunbar on 31 December 1839 by Governor Gawler. A later owner, a Mr Mueller, subdivided the land in 1919. Muller Road adjoins the southern boundary of Greenacres and is a corruption of Mueller.

Greendale *NSW*, a suburb of Liverpool, takes its name from a property originally held by Mary Birch. It was once noted for its timber and wheat.

Greenfield Park *NSW*, is a suburb of Fairfield on the south-western outskirts of the Sydney metropolitan area. It is named after a major road in the area.

Greenhill *SA* is a suburb in the foothills seven kilometres south-east of Adelaide. It takes its name from the colour of the landscape. References to Green Hill and Brown Hill occur early in the history of white settlement in the Adelaide area. Greenhill Rivulet, named by Colonel Light, is now known as First Creek. The Green Hills Special Survey is a different location, south of Echunga, surveyed in the early days of the colony.

Greenmount *WA*, in Mundaring Shire on the outskirts of the Perth metropolitan area, was named in 1827 by Captain James Stirling who viewed the hills from the west and named the first hill Greenmount because of the green tinge.

Greensborough *V* is a locality in the Shire of Diamond Valley on the outskirts of Melbourne. Greensborough takes its name from Edward Bernard Green, a businessman who bought the land in 1841 and later sold allotments in the township.

Greenwich *NSW*, a locality in the Municipality of Lane Cove, bears the same name as the London suburb on the River Thames. The Sydney suburb takes its name from Greenwich House, built in 1836. It stands on the corner of George Street and St Lawrence Street on land originally granted to George Green, a boat builder, in 1836. In 1841 Green announced that he had discovered coal at Greenwich but Governor Gipps refused to allow him to mine it. Later, Captain Gother Kerr Mann (1808-1899) resided there with his large family. Mann called the house 'Willoughby' but the family later changed it to avoid confusion with the postal district of Willoughby.

Grenfell *NSW* is named after T.F. Grenfell, a gold commissioner of the district who was killed by bushrangers at Narromine in 1866.

Greta *NSW* Long before coal was discovered here, a village was surveyed at Anvil Creek in 1842 and named Greta by the

Governor-in-Council after a small river in Cumberland, England. This is probably adjacent to Greta Hall, in Keswick, where there were two houses under one roof. The poets Coleridge and Southey lived one in each house.

Greystanes *NSW*, a suburb of the Municipality of Holroyd, in Sydney's western suburbs, takes its name from the old home of Nelson Lawson, who died in 1849. Greystanes House survived until after World War II when it was demolished.

Griffith *ACT* is named after Sir Samuel Griffith (1845-1920), a former premier of Queensland who was one of the founders of the Constitution.

Griffith *NSW* was designed by Walter Burley Griffin, the American architect who also designed Canberra. The town was named after Arthur Griffith, the Minister of Public Works who, in 1912, opened Burrinjuck Dam, one of the major sources of irrigation water for the area.

Groote Eylandt *NT* means 'big island' in Dutch. The Dutch navigator, Abel Tasman, named it as he passed between the Australian mainland and Groote Eylandt in 1644.

Grose River; Grose Vale; Grose Wold *NSW* After Lieutenant-Governor Francis Grose, who succeeded Governor Phillip. The name was conferred by William Paterson, who explored the Blue Mountains with George Johnston, John Palmer and Edward Laing in 1793. They travelled by boat from the Hawkesbury River. Paterson's objective was to discover a way across the mountains (in which he failed) and to gather botanical specimens for Sir Joseph Banks. His entry was made through the Grose Valley. After Paterson's death his widow married Grose.

Grotto Point *NSW*, a locality in Manly, was first named for the caves found there by the exploratory expedition which Captain Arthur Phillip led to investigate Port Jackson before the First Fleet moved north from Botany Bay in January 1788. Lieutenant W. Bradley used the name in his record of explorations in the harbour on 28 January 1788.

Grovedale *V* is a residential and farming district adjacent to Belmont, Geelong. It was first settled by Lutherans and named Germantown. During the First World War, anti-German hysteria caused a change to the name of an English settler's house.

Guichen Bay *SA* Named by Nicolas Baudin in 1802 after the Dutch Admiral Guichen.

Guildford *NSW*, a suburb of the Municipality of Holroyd, in Sydney's western suburbs, was named after the Earl of

Guildford by his kinsman, Lieutenant Samuel North, who received the original grant in 1837.

Guildford *WA* is in Swan Shire located on the Swan River fourteen kilometres from Perth. Whilst exploring the Swan River in 1827, Captain Stirling was favourably impressed by the fertile alluvial flats around the area now known as Guildford. Returning as Governor of the new colony in 1829 he selected a land grant here, naming it 'Woodbridge' in honour of his wife who was Ellen Mangles of Woodbridge, Surrey, England. The nearby site he reserved for a town he named Guildford after the town of Guildford in Surrey near where he was married in 1828.

Gulf of Carpentaria *NT Q* was probably the first part of Australia reached by Europeans. In 1606, the Dutch ship *Duyfken*, under the command of Willem Jansz, sailed into the gulf as far as Cape Keer-weer (Turnagain). Jansz was followed by Jan Carstenz, who sailed almost to the head of the Gulf in 1623. He named it after Pieter de Carpentier, Governor-General of the Dutch East Indies.

Gulf Saint Vincent *SA* was named by Matthew Flinders on 30 March 1802 'in honour of the noble admiral who presided at the Board of Admiralty when I sailed from England, and had continued to the voyage that countenance and protection of which Earl Spencer had set the example, (Right Honourable John, Earl of Saint Vincent), I named this new inlet the Gulph of St Vincent'. Nicholas Baudin arrived in the Gulf shortly after Flinders and named it Golfe Josephine 'in honour of our august Empress', but Flinders' name took precedence. The Aboriginal name was Wongayerlo meaning 'western sea or lake', or 'overwhelming water where the sun sets'.

Gulgong *NSW* comes from an Aboriginal word meaning 'deep waterhole'.

Gundagai *NSW* comes from the Aboriginal *gunda-bandoobingee*, and is locally shortened to Gundy. The full word means 'cut with tomahawk at the back of the knee' (*gunda*, 'sinews at back of knee' and *bingee*, 'cut with a tomahawk') or 'going upstream'.

Gunnamatta Bay *NSW* in Sutherland Shire was probably named by Lord Audley when he made the first official survey of Port Hacking and the Hacking River in 1863-1864. The Aboriginal meaning was 'a place of beach and sandhills'. It was the original name for Cronulla.

Gunnedah *NSW* This name derives from the Aboriginal root word *guni*, meaning 'white stone'. This accounts for the usual explanation, 'place of the white stone', or 'place of many white stones'. Another conjecture is 'place of the destitute'.

Gunning *NSW* An Aboriginal derivative. Probably a corruption of *goong*, 'swamp mahogany', a tree which has branches of small white flowers and a hard, reddish wood resembling Spanish mahogany.

Gunpowder *Q* is the site of a copper mine about 130 kilometres north of Mount Isa. Its name is taken from the nearby Gunpowder Creek. According to one story, this name was derived from an incident in the district's early history when a number of the Kalkadoon Aboriginals were shot near the banks of this creek. Another story suggests that the name originated when a barrel of gunpowder was accidentally dropped in the creek by an Afghan camel driver.

Guyra *NSW* An Aboriginal word meaning 'fish can be caught'. A pastoral run of this name was held by Charles W. Marsh in 1848. By 1866 there were two properties, Guyra East and Guyra West, but the locality was not proclaimed a village until 1885.

Gwydir River *NSW* Although there have been several conjectures on the meaning of this so-called Aboriginal word, the matter was settled for all time by an entry in Allan Cunningham's diary made when he crossed this river in 1827. 'I named the river after Lord Gwydir.'

Gymea *NSW* in Sutherland Shire is an Aboriginal word for 'giant lily'. It is believed to have been named by W.A.B Greaves, a government surveyor in 1855, after the tall, red-flowered native lily.

Gympie *Q* was the name given by the Kabi Kabi tribe to the stinging tree which grew abundantly beside the Mary River.

Haberfield *NSW*, a suburb of Ashfield, was developed in 1902 from the Dobroyd Estate by two enterprising Summer Hill realtors, Richard Stanton and W.H. Nichols. The Haberfield scheme involved a completely new concept of suburban development. Stanton's idea was based on town planning with covenants to safeguard the single-storeyed garden-suburb character of the district he hoped to create. The project was an immediate success. Today the suburb is known for the Federation architecture of many of its houses.

The Federation style is distinguished by its ornate verandas and the ornaments on the roofs and chimneys.

Hackett *ACT* is called after Sir John Winthrop Hackett (1848-1916) a West Australian newspaper publisher and politician who was a member of the 1897 constitutional committee.

Hackham *SA*, according to one local historian, may have been named after John Barton Hack, a member of the first exploration party to visit the area in 1837. According to Praite and Tolley's *Place Names of South Australia*, the land was bought from James Kingdom by Edward Castle who named it after his English home. Rodney Cockburn's book, *What's In A Name*, says it was named by James Kingdom without leaving any explanation of its derivation. The South Australian Places Names Board has references to all three explanations in its records.

Hahndorf *SA* was named after Captain Dirk Meinhertz Hahn who brought German migrants to South Australia in the ship, *Zebra*, in 1838. The name was changed to Ambleside during the First World War, but reverted to Hahndorf in 1935.

Hall *ACT* recalls Henry Hall (1802-1880) who obtained a land grant which he named 'Charnwood' in 1883.

Hallet *SA*, a town in the mid-north of South Australia, was surveyed in 1870 and proclaimed in 1875. It was named after Alfred Hallett, brother of John, who held a large pastoral station in the area.

Hallett Cove *SA* is a suburb of the City of Marion in the southern part of the Adelaide metropolitan area. The cove was discovered by John Hallett when looking for mission stock in 1837. John Hallett built the first two-storey house in South Australia in July 1837 on South Terrace. In the 1820s Hallett Cove was the site of an unsuccessful goldmining operation by a company formed by John Hallett and his brother, Alfred. The cove is famous because of its geological features which provide evidence of glacial action dating back millions of years.

Halls Creek *WA* is named after Charles Hall who found gold there in 1885 and started the first big gold rush in Western Australia.

Halls Gap *V* is named after C.B. Hall, a pioneer who established cattle runs on the eastern Grampians in 1841. He was the first white man to enter the gap by following the path used by the Aboriginals to cross the Grampians.

Hamersley Range *WA* was discovered by F.T. Gregory in 1861 and named after Edward Hamersley, a supporter of his expedition. Gregory reported the existence of large quantities of iron in the district which is one of the richest mineral areas in the world.

Hamilton *Q* is a locality in the Brisbane metropolitan area and is named after a hotel built by G. Hamilton of Toowoomba.

Hamilton *V* is a city in western Victoria on the banks of the Grange Burn Creek. Major Mitchell named the creek in 1836. The Wedge Brothers named their run 'The Grange' in 1838. The town was surveyed in 1850 by Henry Wade who recommended the name Hamilton after the Duke of Hamilton.

Hamilton Hill *WA* in the City of Cockburn near Perth, was named in 1830 after the farm established there by George Robb.

Hamlyn Heights *V* is a residential suburb in the Shire of Corio. Hamlyn literally means 'dwelling by pool'.

Hammondville *NSW*, a suburb of Liverpool, dates from the depression of the 1930s. Canon R.B.S. Hammond, known for his efforts on behalf of city families in distress through unemployment, was the founder of the settlement that bears his name. The unemployed people housed in Hammondville were flower growers and poultry farmers.

Hampstead Gardens *SA* is a suburb in the corporate city of Enfield in the Adelaide metropolitan area. Hampstead Gardens was named after Hampstead in London.

Hampton *V* is a suburb of the bayside city of Sandringham in the Melbourne metropolitan area. Hampton takes its name from D.B. Hampton, who arrived in the Port Phillip district in 1842. The district was renamed Hampton in 1889.

Harbord *NSW* is a locality in the Municipality of Manly. Freshwater and Harbord began as adjoining localities at the time Lord Carrington became Governor of New South Wales in 1885. However, the growing popularity of surf bathing early in the twentieth century gave beachside Freshwater prominence over its western neighbour, Harbord. In 1923 the whole area was named Harbord (the maiden surname of Lady Carrington). In 1980 the Geographical Names Board re-named Harbord Beach and its adjoining Reserve, Freshwater.

Harcourt *NSW* is a suburb of the Sydney municipality of Canterbury. It was sub-divided in 1889 as Harcourt Model

Suburb, a revolutionary concept in land development. Fifth and Eigthth Avenues were planned as 'grand avenues' with statues and ornamental urns at each corner and elaborate gardens were laid out. Unfortunately the sale coincided with the economic depression of the 1880s and many of the idealistic dreams had to be abandoned. The name was supplied by the developer, W.E.H. Phillips.

Harris Park *NSW*, a suburb of the City of Parramatta, recalls the New South Wales surgeon, John Harris, who bought James Ruse's land and built 'Experiment Farm Cottage'. Harris died at 'Shane Park' in 1838. When his family's property was subdivided in the 1870s, the area became known as Harris Park.

Harrisfield *V* is a suburb of the City of Springvale on the outskirts of Melbourne. It was named during the 1950s after Councillor Harris.

Hartley; Little Hartley; Hartley Vale *NSW* Governor Macquarie named this 'beautiful, extensive vale of five miles' the Vale of Clwydd, in 1815. The township of Hartley was formed and named in 1838 after a place in Northumberland. The name means 'stag hill'.

Hassall *NSW* is a locality in the City of Blacktown on the western outskirts of Sydney. It is named after Rowland Hassall (1768-1820). He was associated with the area as a Church of England Minister at least until 1819. He also acted as agent for the estate of Phillip Parker King and during Samuel Marsden's absences, managed his properties.

Hastings River *NSW* was discovered and named by John Oxley in 1818. Hastings Point is known locally as Cudgera Head, from the Aboriginal word *kudgeree*, 'raw flesh'. It was named after Warren Hastings, Governor-General of India, by John Oxley in 1823. It has, however, been suggested that as Mrs Oxley came from Hastings in England, this may have influenced his choice of the name.

Hawker *ACT* honours Charles Allan Seymour Hawker (1894-1983), a soldier, pastoralist and federal member of parliament.

Hawker *SA* is named after George Charles Hawker, pioneer settler of Bungaree, who achieved a reputation as a breeder and woolgrower, and became Speaker of the South Australian Parliament. It was laid out in 1880 on the Great Northern Railway.

Hawkesbury River *NSW* was named by Governor Arthur Phillip in 1789 after Baron Hawkesbury.

Hawks Nest *NSW* was so named because apparently a large tree near the former hotel was a favourite nesting place for hawks and was used as a navigational marker.

Hawthorn *V* is a city in the Melbourne metropolitan area. The first inhabitants of the area were members of the Wurundjeri tribe who called it Booroondara, which may have meant 'place of shade'. The first recorded visit of white people was made in 1803. And in 1837 John Gardiner and his family settled in Booroondara, as it was known in its early days. The village is supposed to have got its present name because a gentleman named Hawthorne visited the town planner, Robert Hoddle, about 1853 when he was looking for a name for the new village. A careless government printer left off the 'e' when Hawthorn became a municipal district in 1860. Hawthorn was proclaimed a city in 1890.

Hawthorn, Mount *see* **Mount Hawthorn**

Hay *NSW* is named after Sir John Hay (1816-1892), the local member of parliament for Murrumbidgee, who became Speaker of the Assembly in 1862. An earlier name was 'Lang's Crossing' because of the ford on the Murrumbidgee River for cattle bound for Victoria.

Haymarket *NSW* is a locality in the City of Sydney. Haymarket recalls the cattle and corn markets that dated back to 1829 on this site before they moved to Flemington in 1975.

Hazards *T* are solid granite mountains rising almost sheer from the sea. They are named after Captain Albert Hazard, who lost his whaling ship *Promise* in Hazard Bay.

Hazelbrook *NSW*, a locality in the Blue Mountains west of Sydney, got its name when the railway station was opened in 1884. It was probably so named after 'Hazel Brook House', a two-storeyed building close by, which was later demolished.

Healesville *V*, a popular resort township eighty-seven kilometres north-east of Melbourne, was named in honour of Richard Heales, coachbuilder and Premier of Victoria 1860-61.

Heard Island, an Australian overseas territory, is located in the Southern Ocean about 4000 kilometres south-west of Fremantle. The island takes its name from Captain Heard of the American ship *Oriental*, which reached the island in 1853. The nearby McDonald Islands were discovered by Captain McDonald of the British ship *Samarang* in January 1854.

133

Heathcote *NSW*, in Sutherland Shire, was named in 1835 as a civil parish by Surveyor-General Mitchell after one of two fellow-officers who served with him in the Peninsular Wars (1808-1814). The village was originally called Bottle Forest. It was surveyed in 1842 and is now Heathcote East.

Heathcote *V* is a township forty-seven kilometres south-east of Bendigo. The town served the famous McIvor Creek gold-diggings. It was named after Sir William Heathcote, a British MP.

Heatherton *V* is a suburb of Moorabbin, a city in the Melbourne metropolitan area. Heatherton was selected in 1880 by Mrs Peter Hotton (wife of the postmaster) at the request of the local people.

Heathmont *V* is a suburb of the City of Ringwood in the Melbourne metropolitan area. Heathmont is located on a low hill that was presumably once covered with heather.

Heathpool *SA*, a suburb of the corporate city of Kensington and Norwood, was originally occupied by George Reed, a native of the Heathpool in Northumberland, England, who arrived in South Australia in 1840. His widow began to subdivide Reed's 'Heathpool' farm shortly after his death in 1878, but most of the farm was still intact in 1920, and the last subdivision did not occur until 1927.

Hebersham *NSW* is a locality in the City of Blacktown on the western outskirts of Sydney. It received its name in 1829 when the Trustees of the Clergy and School Lands in New South Wales planned to establish a village of that name on the Western Highway, west of Eastern Creek. The name Hebersham was chosen to honour Bishop Heber of Calcutta, whose diocese included the colony of New South Wales.

Hectorville *SA* is a suburb of the corporate city of Campbelltown in the Adelaide metropolitan area. It is named after John Hector, the first manager of the South Australian Savings Bank. Hector held land in this area for a gentleman in England. When he sold a parcel of the land to Patrick Boyce Coglin, he named the new suburb in 1855 after John Hector, who had once owned the land at Glenelg beside the Old Gum Tree where the Proclamation of South Australia was read in 1836. Hector conveyed the land in 1857 to the town of Glenelg in order to preserve the tree as a monument to the founding of the state.

Hedland, Port *see* **Port Hedland**

Heidelberg *V* is an old established residential city on the Yarra River about fourteen kilometres from the centre of

Melbourne. It is known because the group of artists called the Heidelberg School painted the landscape of the district in the 1880s. Two halls of the Heidelberg Town Hall have been named after Arthur Streeton and Charles Conder, two famous painters of the group. Robert Hoddle, the government surveyor from Sydney, completed his survey of Victoria in 1837. He named the Heidelberg district Keelbundora, an Aboriginal name for a swamp. Thomas Walker, a wealthy Sydney merchant and investor, purchased large areas of Heidelberg in the first land sales held in September 1838. He then commissioned Richard Henry Browne to act as his agent to resell the land. In 1839, plans for a village to be called Warringal were submitted. Brown is later credited with naming the village Heidelberg because the Yarra River and the Australian Alps reminded him of the Neckar River and the mountains around Heidelberg in Germany. A road board was set up in 1840. Heidelberg became a district in 1860 and a shire in 1871. It was proclaimed a city in 1934.

Heirisson Island *WA* in the Swan River is named after a French officer who was in command of an expedition that explored the Swan River in June 1801.

Helena Valley *WA* is located on land originally granted to Sir James Stirling in September 1829. The suburb name also applies to the valley through which the Helena River flows, and the river was first explored in October 1829 by Ensign Robert Dale. It is thought to be named by Dale after Helen Barbara Dance, or perhaps after Dale's sister, Helen Catherine. 'Helena' was used, so as not to confuse the river with Ellen Brook.

Hemmant *Q* is a locality in the Brisbane metropolitan area and is named after William Hemmant (1838-1916), a draper and politician. He was Treasurer from 1874 to 1876.

Hen and Chicken Bay *NSW* is a locality in the Sydney municipality of Concord. It is an intriguing name of uncertain origin. It is surmised that Captain Hunter's chart-making boat party in February 1788 gave it that name because a number of sandstone boulders prominent on the shore near to its entrance had a resemblance to the shape of a hen and her chickens. The name, however, does not appear on Hunter's published chart. Another surmise is that Hunter and his party saw an emu and her chick on the shore there.

Henderson *WA* in the City of Cockburn near Perth is named in honour of Admiral Sir Reginald Henderson whose report in 1911 led to the beginning of the Henderson Naval Base at Woodman Point.

Hendon *SA* is a suburb within the Corporation of Woodville, a city in the Adelaide metropolitan area. Hendon was named by Captain Harry Butler, who established an aerodrome there in 1920. He called it after the famous aerodrome in England. Butler sold a portion of his estate to Wilkinson Sands and Wyles Ltd., who subdivided and called the area Hendon in 1921. The aerodrome was later bought by the Commonwealth and became Adelaide's first official airport. Adelaide airport at West Beach is now the city's international airport. Subdivision of the suburb of Hendon during the 1920s made landing dangerous and Parafield became the new airport in 1927. Some streets are named after well-known planes of the time—Avro, Sopwith, Farman, De Havilland and Vickers.

Henley *NSW*, a locality in the Municipality of Hunters Hill, was named by nostalgic Englishmen when rowing became popular on the Parramatta River, as it was on the Thames. The first regatta was held in 1847.

Henley Beach *SA* is a seaside resort in the western area of the Adelaide metropolitan area. It gets its name from Henley-on-Thames, near London.

Henry, Point *see* **Point Henry**

Hepburn *V,* the site of mineral springs north-west of Daylesford, is named after Captain Hepburn who was among the first overlanders who brought cattle from Sydney to Port Phillip. The Aboriginal name was Morrekyle.

Herbert River; Herbert Range *Q* were discovered by G.A.F.E. Dalrymple in 1864, and named by A.C. Gregory, the Surveyor-General, after Sir Wyndham Herbert, the first premier of Queensland.

Hermannsburg *NT* was established as an Aboriginal mission by three Lutheran missionaries in 1877. They named the site 'Hermannsburg' after the Hermannsburg Mission Institute, Germany.

Herne Hill *V* is part of the City of Newtown, Geelong. It was called after a locality of the same name in England. The name comes from the Saxon *hyrne*, a 'nook' or 'corner'.

Heron Island *Q* is a tourist resort of about seventy kilometres from Gladstone. It was probably named by Captain Blackwood, who saw a number of herons when he visited the island while mapping the Great Barrier Reef in HMS *Fly* between 1842 and 1846.

Hervey Bay *Q* was named by Captain Cook after the surgeon who sailed with him on the *Endeavour* in 1770.

Hexham *NSW* After the market town in Northumberland, on the River Tyne.

Heysen Trail *SA* is a long-distance walking-track. Its name honours Sir Hans Heysen (1877-1968), a German born painter known for his paintings of gum trees and scenes in the Flinders Ranges of South Australia. The 800 kilometre horse-riding and walking trail from Mount Babbage (Mount Hopeless) in the Flinders Ranges to Cape Jervis on the south coast of Adelaide (Fleurieu Peninsula) was announced in 1970. It was proposed to link as many national parks as possible with the trail.

Heywood *V* was originally called 'The Edgars' after David Edgar, who built a hotel, the 'Bush Tavern', on the site of the present hotel in 1842, but the name was later changed to Heywood.

Hicks, Point *see* **Point Hicks**

Higgins *ACT* is named after Henry Bournes Higgins (1851-1929) who was a judge of the High Court from 1906 to 1929.

Highbury *SA* is a locality in the City of Tea Tree Gully in the north-eastern section of the Adelaide metropolitan area. It was the name of the property of Stephen George Dorday, who bought a farm there in 1850. Highbury is named after a town in Middlesex, England.

Highett *V* is a suburb of the City of Moorabbin in the Melbourne metropolitan area. Highett was named after William Highett, a banker who owned a cattle-run near Hampton.

Highgate *WA* is a suburb which was known in 1884 as Highgate Hill, a name presumed to be coined by Surveyor Charles Crossland after his home town in England. Highgate Hill's major landmark, St Albans Church, was opened on 13 June 1889, four days before St Albans Day when it was first used for the feast of St Albans. St Albans Cathedral is in Hertfordshire, England.

Highton *V* is a residential locality, south-west of the City of Geelong, in Barrabool Hills. John Highett built a house on the highest position in the area which became known as Highett's town.

Hillcrest *SA* is a suburb in the corporate city of Enfield in the Adelaide metropolitan area. Hillcrest is a descriptive name.

Hillsborough *NSW*, a Newcastle suburb, is named after the owner of the original subdivision, Dr J.J. Hill, who was Mayor of Lambton and a large shareholder in the colliery company as well as medical officer of the local collieries.

Hillsdale *NSW* is a suburb of the Sydney municipality of Botany. It was originally called Matraville, but it was changed in 1961 to avoid confusion with the area of the same name across the border in Randwick. The new name honours Pat Hills, a Labor Minister of State Parliament. Another politician commemorated in the area is the former premier of New South Wales, John Joseph Cahill, who has a high school named after him. It was opened in 1961 by his wife.

Hillston *NSW* After William Hill, who built and owned a hotel there. The locality was first known as Redbank, and by the Aboriginal name *melnummi*, 'red soil'.

Hilton *SA*, according to the South Australian nomenclature expert, the late Rodney Cockburn, is named after Matthew Davenport Hill QC (Recorder of Birmingham) who was the original holder of the section. A local hotel called the Hilton Hotel causes some confusion among visiting American tourists as they drive past the humble hostelry named after the suburb on their way from the airport to Conrad Hilton's much grander hotel in the city of Adelaide.

Hilton *WA* was named after Hilton Park, a reserve in the area. The name, Hilton, was suggested by the Postmaster-General's office in May 1964. The locality was shown as Hilton Park on some maps until May 1959 when it was agreed that the generic term 'Park' should not be used.

Hinchinbrook Island *Q* is the world's largest maritime island national park. James Cook named it in 1770. It is called after the family seat of his patron, George Montagu Dunk, who was First Lord of the Admiralty.

Hindmarsh *SA* is both an inner suburban area of Adelaide and the name of a corporation which has several other suburbs under its control. It was the first private township formed in South Australia. The suburb takes its name from John Hindmarsh, the first governor of South Australia, who arrived when the colony was founded in 1836. He acquired the site of the future Hindmarsh Village in May 1838, shortly before he was recalled to England. He subdivided the land into two-hundred half-acre lots. The first purchasers included men now remembered by streets named after them, although none of them appear to have lived in Hindmarsh. The name of the village was chosen by vote. There were one hundred and sixty-two votes for Hindmarsh, thirty-three for Victoria, four for Wakefield and one for Dawlish. It was not recorded how many votes were cast by actual land purchasers and how many were the result of proxy votes by Governor Hindmarsh and his nominees.

Hindmarsh, Lake *see* **Lake Hindmarsh**

Hobart *T* is named after Robert Hobart, 4th Earl of Buckinghamshire and Secretary of State for War and the Colonies from 1801 to 1804. Lieutenant John Bowen had formed a settlement at Risdon Cove, but Lieutenant-Governor David Collins considered the site unsuitable, and shifted it to Sullivan Cove, taking possession on 19 February 1804. Bowen had already called the Risdon Cove settlement Hobart in a dispatch to the Governor of New South Wales dated 27 September 1803, bearing this address. Collins transferred the name as well as the settlement to Sullivan Cove. Early forms of the name were Hobart Town, Hobarton, and Hobarttown. In 1813, Lieutenant-Governor Davey tried to have Hobart Town shortened to Hobart, but official approval was not given until 1881.

Hobart Street Names Governor Lachlan Macquarie named most of Hobart's streets during his visit there in 1811. He recorded in his journal on 2 December 1811: 'I had the names of the great square and principal streets painted on boards and this morning erected on posts at the angles . . . named them as follows: viz., George's Square, 1 Macquarie (main) Street, Liverpool Street, Argyle Street, Elizabeth Street, Murray Street, Harrington Street and Collins Street'.

He gave no reasons for so naming them, but George's Square, which was later dispensed with, was presumably named after King George III; Macquarie Street must have been named after himself; Liverpool Street was undoubtedly named after the Earl of Liverpool, then Secretary of State for the Colonies; Argyle Street was probably, like Argyle Plains in Nerw South Wales, named after the Duke of Argyll, head of the Clan Campbell; Elizabeth Street was called after his wife, Elizabeth Henrietta, née Campbell; Murray Street was named after Captain John Murray, Commandant at the Derwent, 1810; Collins Street was named after Lieutenant-Governor David Collins, the founder of Hobart Town. A plaque affixed by the City Council states that Harrington Street was named after the Earl of Harrington, who was Colonel of the Life Guards, but this has not been proved. A street plan of Hobart Town executed by order of Governor Macquarie in 1811 also shows the street now known as Davey Street was then known as Pitt Street, presumably after William Pitt the Younger. Thomas Davey was the second Lieutenant-Governor of Van Diemens Land. It is not recorded when Davey Street was given this name.

Hobsons Bay *V*, the northern extremity of Port Phillip Bay, was named by Sir Richard Bourke in 1837 on a trip to Port

Phillip Bay. The name honours Captain William Hobson, commander of HMS *Rattlesnake* who surveyed the bay.

Hoddle *V* a small rural locality south-west of Foster, was named after Robert Hoddle, Victoria's first Surveyor-General.

Holbrook *NSW* was named after Lieutenant Norman Holbrook, RN, who was awarded the Victoria Cross in World War I for taking his submarine underneath mines to torpedo the Turkish battleship *Messudiyeh*. The first name, given in 1858, was Ten Mile Creek. This was replaced by Germanton in 1875, in honour of the first settler, John Pabst. During World War I, when anything of German origin was suspect, the present name was substituted.

Holden Hill *SA* is a suburb in the City of Tea Tree Gully in the Adelaide metropolitan area. The name is a corruption of the surname 'Halden'. Robert Halden acquired the land in the 1840s (as late as 1935 the area was referred to in district council minutes as Halden Hill).

Holder *ACT* is called after Sir Frederick William Holder (1850-1909), first speaker of the House of Representatives from 1902 to 1909.

Holroyd *NSW*, a municipality in the western metropolitan area of Sydney, is named after its first mayor, Arthur Todd Holroyd (1806-1887). This remarkable man was born in London. He studied medicine but after practising for only a year, he turned to law instead. From 1835 to 1838, Holroyd travelled through Egypt, the Sudan and Syria. On his return to London, he exposed the horrors of the slave trade. In 1843, he migrated to New Zealand. In 1845, he arrived in Sydney where he became one of the colony's leading barristers. Holroyd had many interests. He is remembered as the father of lawn bowls in Australia. He was also a director of the Australian Mutual Provident Society and the Ophir Gold Mining Company. In 1855 he bought a large portion of William Sherwin's estate and named it 'Sherwood Scrubs'. Here he established the Sherwood Drain and Tile works. In 1872 he was largely responsible for establishing the Municipality of Prospect and Sherwood and was its first mayor. The name was changed in 1927.

Holsworthy *NSW*, a suburb of Liverpool, is named after a village in Devon, England. Lord Kitchener visited the area in 1910 and since 1913 it has been known for its military establishments.

Holt *ACT* honours Harold Holt (1908-1967) who was Prime Minister of Australia from 1966 to 1967, when he disappeared while swimming at Portsea near Melbourne.

Holtmere *NSW,* in the Kurnell area of Sutherland Shire, covers an area of land originally held by Thomas Holt. It appeared on the first map of the shire.

Home Hill *Q* is a town on the Burdekin River about eighty kilometres south of Townsville. It came into being in 1911 when the township was designed by the surveyor, R.A. Suter, and surveyed by Alfred Marshall in the same year. In the early stages the township was referred to as Inkerman, but it was officially named Home Hill by the Surveyor-General. The story goes that it was to have been known as Holme Hill, after a battle in the Crimean War, but the young man sent to paint the name on the railway station painted 'Home Hill', and so it has remained.

Homebush *NSW* is a suburb of Strathfield Municipality. Although properly the name of D'Arcy Wentworth's estate, it was applied loosely to the small settlement on both sides of Sydney Road until its first official usage as the Village of Homebush after subdivision of the Underwood property in 1878. There is a belief that Homebush was so named as it was the last bush area in which bullock teams were rested before making the final day's journey into Sydney Town—literally the Home Bush, the bush near home. However, no evidence has ever been produced to support this story which appears to have more weight in fiction than in fact.

Hope Valley *SA* is a suburb in the City of Tea Tree Gully in the north-eastern section of the Adelaide metropolitan area. Hope Valley began to emerge as a township as early as 1842. It was named in 1842 by William Holden, a former journalist who opened a butcher's shop and store there. The store was burned to the ground, but far from feeling downcast by these misfortunes, Holden felt himself 'inspired by hope', hence the name. William Holden left the district in 1851 after his wife, Sally, was killed in a riding accident. He later worked as a journalist on the *Register*. Douglas Tolley, who bought land at Hope Valley in 1891, began a vineyard as a hobby and saw it grow to a large business before his death in 1932.

Hopeless, Mount *see* **Mount Hopeless**

Hopetoun *V,* a township in the Mallee, by Yarriambiac Creek, was named after the Earl of Hopetoun, Governor of Victoria. It was once known as Lake Coorong, after the lake by that name.

Hornsby *NSW,* a shire on the north-western outskirts of Sydney, gets its name from Hornsby Place, the grant given to Samuel Horne, who was sentenced to transportation for life at the age of nineteen. He was given a full pardon in 1831 for his part in helping Constable Thorn to capture the bushranger, McNamara. His grant of land was actually south of the present Hornsby area in the district now called Normanhurst.

Horsham *V*, a town on the Wimmera River in Western Victoria, was named by the first settler, James Monckton Darlot, from his home town in Sussex after he had established himself here in the 1840s. A post office and store began operations in 1849 and the town was surveyed in 1854. The Aboriginal name was Bongambilor meaning 'place of flowers'.

Horsley Park *NSW* is a suburb of Fairfield on the south-western outskirts of the Sydney metropolitan area. Horsley Park is an area which Governor King awarded to Colonel Johnston in 1804 as a reward for putting down a convict rebellion. This became known as 'King's Gift'. On Johnston's death, the property was willed to his daughter Blanche, who in 1829 married Major Weston of the Indian Army. When Weston built a homestead on the property, he changed the name to 'Horsley', after his birthplace in Surrey, England. The suburb of Horsley Park gets its name from the house.

Hotham, Mount *see* **Mount Hotham**

Houghton *SA* is a locality in the City of Tea Tree Gully in the north-eastern section of the Adelaide metropolitan area. Houghton was one of the earliest settlements in the Adelaide Hills. It was laid out as a village in 1841 by John Richardson, a land agent and surveyor, who had arrived in the colony in 1838. He bought land from John Barton Hack and placed young Joseph Barritt in charge of his property, which he called 'Houghton Lodge'. A village developed at Houghton in the 1840s, but because the main route from Adelaide later by-passed the village it did not develop as rapidly as its neighbours, which were founded later. Houghton is named after Houghton-le-Spring in Durham.

Houtman Abrolhos *WA* is an eighty-kilometre chain of reefs and islands about sixty kilometres off the coast of Geraldton. The islands were named after a Dutch captain Frederik de Houtman who discovered them in 1619. *Abrolhos* is Portuguese for 'open your eyes' which is an appropriate warning in view of the number of ships that have been wrecked there.

Howlong *NSW* A corruption of an Aboriginal name, Oolong, 'place of native companions'.

Howrah *T* was named after 'Howrah House', a property established in the 1830s on Clarence Plains by a retired Indian Army officer, who took the name Howrah from a place of the same name near Calcutta.

Hoxton Park *NSW*, a suburb of Liverpool, was originally known as Cabramatta or Sarahville and the local sawmills were an important source of timber. In 1887 a land syndicate bought much of the area and renamed it Hoxton after the property of Thomas S. Amos, a lawyer who was granted land in 1816. His land grant adjoined that of Judge Barron Field.

Hughenden *Q* is a town on the Flinders River 1720 kilometres north-west of Brisbane. Ernest Henry took up land there in 1864. He named 'Hughenden' station after Hughenden Manor, the home of his grandfather in Buckinghamshire, England.

Hughes *ACT* is named after William Morris Hughes (1864-1952), Prime Minister of Australia from 1915 to 1923.

Hughesdale *V* is a locality in the City of Caulfield in the Melbourne metropolitan area. Hughesdale was named in honour of Mayor James V. Hughes of Oakdale by the Victorian Railways in 1925.

Hume *ACT* honours Hamilton Hume (1797-1873), the explorer who pioneered the overland route between Sydney and Port Phillip.

Hume, Lake *see* **Lake Hume**

Humpty Doo *NT* According to Bill Beatty in *Unique to Australia*, the first white settler made occasional trips to Darwin, and when asked how he was getting on, replied 'Everything's humpty doo'.

Humpybong *Q* is an Aboriginal name believed to mean 'dead' or 'empty dwelling'.

Hunter River *NSW* was named after Governor (and Admiral) John Hunter, by Lieutenant T.G. Shortland in 1797. It had previously been discovered by a party of escaped convicts, in search of whom Shortland had been dispatched by the Governor. A fishing party had found coal at Newcastle, and Shortland made a similar find at the river which for some years was known as Coal River. The Aboriginal name was either Coonanbarra or Gognon (or both).

Hunters Bay *NSW*, a locality in Mosman on Sydney's North Shore in which Balmoral Beach is located, is named after Captain John Hunter of the *Sirius* and the second governor of New South Wales.

Hunters Hill *NSW* commemorates Captain John Hunter, the officer who commanded the *Sirius* and later made the first survey of Sydney Harbour. In the early days of settlement the name was applied to the whole North Shore area of the harbour. By coincidence, one of the Scottish Martyrs, Thomas Muir, bought a farm on the North Shore (probably around the present site of Kirribilli Point) which he named 'Huntershill' after his father's farm at Hunters Hill, Glasgow. The large district of Hunters Hill was gradually broken into smaller parts which were renamed Willoughby, Ryde and so on. The Municipality of Hunters Hill, incorporated in 1861, preserved the historic name. The Aboriginals called the peninsula Moocooboola, 'the meeting of the waters' and the name has been preserved on the Municipality's coat of arms if not the map.

Huntleys Point *NSW* is a locality in the Sydney municipality of Hunters Hill. Huntleys Point recalls A.R. Huntley who bought land from the Crown in 1851. 'Point House', which he built there, still stands at number 22 Huntleys Point Road. It was extended in 1880 from five rooms to its present size. His son, Stafford Huntley, lived in the house until the 1960s.

Huon River *T* was named by D'Entrecasteaux after his second-in-command, Captain Huon de Kermadec.

Hurlstone Park *NSW* is a suburb of the Sydney municipality of Canterbury. It is named after the Hurlstone College, founded in 1878 by John Kinloch on the site of Trinity Grammar School (Hurlstone was his mother's maiden name). In 1911 the nearby suburb of Fern Hill was changed by popular vote to Hurlstone and the railways department added Park to avoid confusion with the railway station, Hillston, in western New South Wales.

Hurstbridge *V* is a locality in the Shire of Eltham on the outskirts of Melbourne. Hurstbridge was originally called Hurst's Bridge. It takes its name from the Hurst family. Henry Hurst arrived in Melbourne in 1853 and first came to the Eltham district about 1859.

Hurstville *NSW* is a city on the Georges River to the south of Sydney. Hurstville was suggested by a school inspector, Mr McIntyre, when consideration was being given to the establishment of a public school in 1876. It is a combination of *hurst*, 'a wooded eminence' and *ville*, 'town'. The township was previously known as Lord's Forest, then Gannon's Forest. The post office was established as Gannon's Forest and the name changed to Hurstville in 1881. When the railway was built in 1884, the station was named Hurstville.

Hutt River *WA* was named by the explorer, George Grey, on 5 April 1839 after William Hutt M.P., brother of the Governor of Western Australia, John Hutt. In 1970, a wheat farmer named Leonard Casley put the area on the map after he had a difference of opinion with the Western Australian Government over wheat quotas. He formally seceded from the state, and declared Hutt River province to be an independent province under the British Crown. He adopted the title of Prince Leonard. The breakaway province issued its own stamps and money, but it has never been recognised by the Australian Government or by any other country.

Illawarra *NSW* An adaptation of an Aboriginal word, Elouera, Eloura, or Allowrie, variously translated as 'pleasant place near the sea', or 'high place near the sea', or 'white clay mountain'. *Wurra* or *warra* probably means 'mountain', and *illa* may be 'white clay'. George Bass and Matthew Flinders were the first to visit the district. Flinders recorded the fact that it 'was called Alowrie by the natives'. It was first known as Five Islands, but the name of the group of five islands off Red Point was transferred to this part of the coastal district, seventy kilometres south of Sydney, later known by the Aboriginal name of Illawarra. It was in use in 1806. In 1817 Governor Macquarie wrote of 'part of the coast known generally by the name of the Five Islands, but called by the natives Illawarra'.

Illawarra, Lake *see* **Lake Illawarra**

Illawong *NSW*, in Sutherland Shire, was known as East Menai but changed to Illawong when the public school was erected in 1960. It is an Aboriginal word meaning 'between two waters'.

Iluka *NSW* An Aboriginal name meaning 'near the sea'.

Indooroopilly *Q* is a locality in the Brisbane metropolitan area and is an Aboriginal word meaning 'running water' or 'gully of leeches'.

Ingham *Q* is called after William Bairstow Ingham who settled in the district in 1873 and founded Ings plantation. The town was named in his honour in 1882.

Ingle Farm *SA* is a suburb in the City of Salisbury. According to one local historian it was named by Jabez Rowe, a grandson of James Rowe who first went to Kangaroo Island in 1836 and took up land in the Salisbury district in 1849. According to this historian, Jabez named the farm as a result of his marriage to a Miss Wright of Inglewood. The South Australian Place Names Board has no record of this story. The board says that Jabez Rowe was the registered proprietor of land in this area from about 1902. The first recorded subdivision referred to as 'Ingle Farm' was dated August 1965.

Ingleburn *NSW*, a suburb of Liverpool, is named after the Gaelic words for 'bend in the creek'.

Ingleside *NSW* is a locality in Warringah Shire. Ingleside takes its name from the name of a mansion built by a German confidence trickster named Carl von Bieren who described himself as 'the gunpowder king' and who arrived in the district in 1884. He gave a special party for invited guests to see a demonstration of the first Australian-made gunpowder. When it failed to ignite, von Bieren sent an employee to check the fuse. He arrived just as the gunpowder exploded. Where the gunpowder came from is a mystery, because von Bieren never actually produced any gunpowder. After attempts to raise money to finance the non-existent gunpowder manufacturing enterprise, von Bieren left Australia under the name of Walbridge. The police managed to find a faster ship and met him when his vessel docked in England. Von Bieren-Walbridge was returned to Australia and served a prison sentence for fraud. Today the name Powder Works Road recalls his enterprise.

Inglewood *V*, a township forty-five kilometres north-west of Bendigo, is named after Inglewood Forest in Cumbria, England. Gold was discovered there in 1859.

Inglewood *WA* is a suburb of Stirling City in the Perth metropolitan area. The name comes from the Inglewood Estate, the name given to a residential subdivision by the developing company in 1895.

Innaloo *WA* in Stirling, a city in the Perth metropolitan area, was originally called Njookenbooroo when it was developed by settlers in the early 1900s, but the original name of the area was changed to Innaloo when large-scale State Housing Commission development was implemented in the years following World War II. Innaloo was approved as the name for the locality in 1927, being taken from a list of names compiled by Daisy Bates in 1913. Innaloo is the name of an Aboriginal woman of the Dongara region.

Innamincka *SA* A corruption of the Aboriginal word *yidniminkani*, which means 'you go into the hole there'. This commemorated an event when a totemic ancestor commanded a crocodile to disappear into a hole. The townsite was named Hopetoun, but local preference for the Aboriginal name won the day. The name was changed from Hopetoun to Innamincka on 28 January 1892. Hopetoun was a former governor of Victoria, who later became Australia's first governor-general.

Innisfail *Q* was originally called Geraldton after Thomas Henry Fitzgerald (1824-1888), a sugar pioneer, but in 1910 it was changed to Innisfail. This was done to avoid confusion with Geraldton in Western Australia.

Inscription Point *NSW* in Sutherland Shire at Kurnell, was named by the Australasian Philosophical Society in 1822 after they had affixed a plaque to the cliff-face where the *Endeavour* crew first landed.

Inverell *NSW* was a name given by Alexander Campbell to his estate in 1848. The word is derived from Gaelic and means 'meeting of the swans'. The MacIntyre and Swanbrook Rivers meet here and Campbell was impressed by the many swans to be seen on the two rivers. The locality was once known as Green Swamp while the Aboriginal name was Giree Giree meaning 'river with high banks'. The town was surveyed in 1855 and a municipality was proclaimed in 1872.

Ipswich *Q* was originally called Limestone because Captain Logan found limestone there in December 1826. Some months later he sent an overseer and five convicts to quarry the limestone and to erect a lime-burning kiln. The lime was required for the erection of the stone buildings of the settlement of Brisbane. It was burned weekly and taken to Brisbane in small boats down Bremer's Creek and the Brisbane River.

It was at first intended to retain the original name of Limestone for the town. Governor Gipps, who had previously visited the site and had decided that a town should be established there, decided to adopt the name of the town of Ipswich on the River Orwell in Suffolk, England, for the new town. He gave orders to E. Deas Thompson, colonial secretary in Sydney, to write to the Surveyor-General on the subject. There is no record of the Governor's reason for choosing the name.

Iron Knob *SA* So named from the rich deposits of iron ore in this hill and the adjacent Iron Monarch. The towns were built when the Broken Hill Proprietary began the manufacture of steel in 1911. Iron Knob was proclaimed a town on 11 February 1915. Iron Monarch, which is in the locality of Iron Knob, takes its name from the same source.

Irvine, Mount *see* **Mount Irvine**

Isa, Mount *see* **Mount Isa**

Isaacs *ACT* is called after the former judge of the High Court, Sir Isaac (Alfred) Isaacs (1885-1948), who served as the first Australian-born governor-general from 1931 to 1936.

Isabella Plains *ACT* were named after Isabella Maria Brisbane (1821-1849), daughter of Sir Thomas Brisbane, Governor of New South Wales 1821-1825.

Ivanhoe *NSW* Adopted from Sir Walter Scott's novel.

Ivanhoe *V* is a suburb of the City of Heidelberg in the Melbourne metropolitan area. Ivanhoe was named after the Ivanhoe Estate which belonged to Richard Pennor. The name was in use by 1853 and comes originally from Sir Walter Scott's novel *Ivanhoe*.

Jackson, Port *see* **Port Jackson**

Jandakot *WA*, in the City of Cockburn near Perth, comes from an Aboriginal word said to mean 'place of the whistling eagle'.

Jannali *NSW*, in Sutherland Shire, is an Aboriginal word meaning 'moon'. The name was given when the railway station was opened in 1931.

Jenolan Caves *NSW* are limestone caves in the Main Range, about 160 kilometres by road, west of Sydney. 'Jenolan' was the Aboriginal name of a nearby mountain with the meaning of 'a high place'. It was officially adopted for the caves in 1884. At first they were known as McKeown's Caves after an escaped convict named McKeown (or McEwan). According to tradition, James Whalan tracked McKeown into the caverns. However no authentic record of such an escaped convict has been found. The stream that flows through the Devil's Coach House still bears McKeown's name. Charles Whalan (brother of James) may have been the first to explore the caves. Originally the caves were known as the Fish River Caves. The Aboriginal name was Binoomea, meaning 'holes in a hill'.

Jeparit *V*, a township on the shores of Lake Hindmarsh, comes from an Aboriginal word meaning 'small bird'. Jeparit was the birthplace of Sir Robert Menzies, Prime Minister of Australia from 1939-41 and 1949-65.

Jericho *T* During the years from 1804 to 1809, Hugh Germaine, a private of Marines, and two convicts ranged widely through the hinterland of Hobart Town shooting kangaroos which provided food for the settlement. Germaine was reputed to have two books in his saddlebag, the *Bible* and the *Arabian Nights*, and these constant companions suggested names such as Jericho, Jerusalem (now Colebrook), River Jordan, Lake Tiberias, Baghdad, Abyssinia etc.

Jerilderie *NSW* An Aboriginal word meaning 'reedy place', possibly derived from Jereeldrie. There is an amusing legend which has no basis in fact, that the wife of an early settler called her husband 'Gerald dearie' and so the settlement received its name.

Jervis Bay *NSW* was first entered by the transport *Atlantic* in 1791 and named by Lieutenant Richard Bowen, the naval agent on board, in honour of Admiral Sir John Jervis, afterwards Earl St Vincent, under whom he had served.

The Australian Broadcasting Corporation favours the pronunciation Jervis (rhymes with service) rather than Jarvis, on the grounds that this is the pronunciation used by the majority of people, although either pronunciation is acceptable.

Jibbon *NSW* is a locality in Port Hacking on the southern outskirts of the Sydney metropolitan area. Its origin is uncertain. The earliest and, in fact, the only reference to it occurs in the *Trigonometric Survey of New Sales Wales Register of Stations*, 1895.

Jindabyne *NSW* was first known as Jindaboine, and appears in some of the official maps spelt in this manner. It is an Aboriginal word. One very old resident told Snowy Mountains historian George Petersen that the Aboriginals named it after the feather tailed rat. However, the official meaning of the word as recognised by the Australian Museum is 'valley'. Other old settlers also told Petersen that there was a fuller meaning of 'the valley of the waters'.

John Forrest National Park *WA* is located on the Darling Range about twenty-six kilometres east of Perth. It was named in 1947 after Lord Forrest, one of Western Australia's greatest statesmen and explorers.

Jolimont *V* is part of East Melbourne. Governor Charles La

Trobe named his cottage 'Jolimont' after his father-in-law's Swiss home.

Jolimont *WA* was taken from Jolimont Terrace. The subdivision of land by J.L. D'Arcy Irvine in October 1892 saw one road in the subdivision being named Jolimont Terrace, possibly after his home town in Victoria.

Jordanville *V* is a suburb of the City of Waverley in the Melbourne metropolitan area. Jordanville almost certainly honours John Jordan, one of the first men to be granted land in the district and a pioneer of local government in the area.

Jukes, Mount *see* **Mount Jukes**

Julia Creek *Q* Named by R. O'H. Burke after an actress, Julia Matthews.

Jumbuk *V* is a rural district south of Morwell. *Jumbuk* is an Aboriginal word meaning 'cloud' or 'sheep'. Because of the white fleeces of the sheep the Aboriginals applied the word for clouds to them, and the early settlers adopted the term.

Junee *NSW* An Aboriginal word meaning 'speak to me'. An earlier name was Loftus. The present name was first spelt Jewnee.

Kadina *SA* A corruption of the Aboriginal name, Caddy-inna or Caddy-yeena, which means 'lizard plain'. The township was surveyed in 1861, and named by Governor MacDonnell after the locality so named by the Aboriginals.

Kakadu National Park *NT* takes its name from the original people in the region, the Gagudju.

Kalamunda *WA* is a shire in the Perth metropolitan area. When the town-site was approved in 1901, no Aboriginal name for the area was known, so a committee of residents searched through Bishop Salvado's dictionary of Aboriginal words to find a name meaning 'home in the bush'. They decided on *cala*, 'bush' and *munnda*, 'hearth'. The settlers dropped the second 'n' and the Surveyor-General officially changed the initial letter to 'K' to bring it into line with the Royal Geographical Society's system of spelling Aboriginal names.

Kaleen *ACT*, meaning 'water', comes from the language of the Wiradhuri Aboriginal tribe of the Central West of New South Wales.

Kalgoorlie *WA* is believed to have been named from an Aboriginal word *kalgoolah*, which was their name for the silky pear which produces an oval-shaped fruit and which they used to eat raw or roasted. The settlement was originally called Hannans, after Paddy Hannan who began a huge gold rush when he discovered gold there in 1893. The town was named Kalgoorlie when it was proclaimed in 1895.

Kallista *V* is a locality of the Shire of Sherbrook on the outskirts of Melbourne. *Kallista* means 'most beautiful' in Greek. Mrs Eastaugh, a teacher of languages, won the prize for the name in a contest held in the 1920s. It was formerly known as South Sassafrass.

Kambah *ACT* was the name of a former homestead in the Tuggeranong district.

Kambalda *WA* is a mining settlement. In 1897, gold was found by the shore of Lake Lefroy some thirty kilometres north of Widgiemooltha. The town that sprang up was referred to as Redhill, but was gazetted as Kambalda in December 1897. By 1906, the gold had run out and the town forgotten. In 1954, nickel samples were taken there by a prospector named George Cowcill. Ten years later, Cowcill and his partner resurrected their samples and took them to Western Mining Corporation who gave them a reward of $50 000. By 1967 nickel mines and treatment plant were in operation. The new town-site is about three kilometres north of the mining centre. It is gazetted as Kambalda East and Kambalda West. The meaning of the name is unknown.

Kameruka *NSW* An Aboriginal word meaning 'wait here till I arrive'. The original run of this name was taken up by W. and J. Walker.

Kangaroo Ground *V* is a locality in the Shire of Eltham on the outskirts of Melbourne. Kangaroo Ground was so named by the early settlers because of the kangaroos that gathered there.

Kangaroo Island *SA* was named by Matthew Flinders during his visit there in 1802. He chose the name because the kangaroos he found there gave his men a fresh supply of meat. Flinders spelt it Kanguroo Island. The Aboriginal name for it was Karta. Baudin's log showed that he named it Isle Borda. Baudin died on the return voyage and journals and charts published by Freycinet and Peron named the island as Ile Decrés after Denis Decrés, a French admiral.

Kapunda *SA* became Australia's first mining town. C.S. Bagot, the young son of Captain Charles Bagot, one of the first men to take up land in the district, made the first copper find while gathering wildflowers in 1842. The first ore was raised in 1844. The town which was laid out shortly afterwards took its name from the mine. The name may derive from the Aboriginal words *kappie oonda*, probably meaning 'water jump out', 'spring', or possibly 'place of smoke', although this derivation is not very likely.

Kaputar, Mount *see* **Mount Kaputar**

Kareela *NSW* in Sutherland Shire was named by the Geographical Names Board, New South Wales Lands Department, in 1968. It is an Aboriginal word meaning 'place of trees and water' or alternatively, 'south wind'.

Karragullen *WA* is a locality near Armadale, a town on the outskirts of the Perth metropolitan area. The name comes from a local Aboriginal word meaning 'red gully'.

Karratha *WA* was gazetted as a town-site on 8 August 1969. The name was taken from the pastoral station from which the land was resumed. The station was named by Mr Baynton or Harry Whittal Venn, the first owners of the property, between 1866 and 1879. This name is of Aboriginal origin. It means 'good country' or 'soft earth'.

Karrinyup *WA* takes its name from Careniup Swamp which was recorded in 1844 by P. Chauncey during a survey of the area. A golf course was established in the 1920s and the foundation committee, which was looking for a name for the course, used an altered form of the name of the nearby swamp Careniup and also applied this name to another small, open swamp in the area. Thus Lake Karrinyup and the Lake Karrinyup Country Club was born, and the club was registered as the holder of the land on 18 June 1929. For many years the golf course was the only occupier of the area. The present suburb of Karrinyup began to develop after 1957.

Karuah; Karuah River *NSW* is an Aboriginal name which has survived in spite of the fact that when Governor Macquarie visited Port Stephens in 1812 he conferred the name Clyde on the river.

Katanning *WA*, a town on the outskirts of the Perth metropolitan area, is probably derived from an Aboriginal word meaning 'chief meeting place'. The Aborigines used to meet at a pool of fresh water in what is now the main park of the centre of town. Some residents, however, maintain it comes from an Aboriginal woman named Kate Ann or Anning. The town was laid out in 1898.

Katherine *NT* takes its name from the nearby Katherine River, which was named in 1862 by John McDouall Stuart. He called it after Catherine Chambers, the daughter of a South Australian pastoralist, James Chambers, who sponsored three of Stuart's journeys to the Northern Territory.

Katoomba *NSW*, a locality in the Blue Mountains west of Sydney, was called Crushers when a railway station was first established there in 1876. It was named after a quarry in the vicinity. The name was changed in 1877 to Katoomba, which is a derivation of *kadumba*, Aboriginal for 'falling water' or 'falling together of many streams'.

Keer-Weer, Cape *see* **Cape Keer-Weer**

Keilor *V* is a city sixteen kilometres north-west of the centre of Melbourne. Some authorities claim Keilor comes from an Aboriginal word for 'brackish water'. Others say it was bestowed by James Watson, the son of a famous Scottish agriculturist whose property on the chapel at Keilor Forfarshire was famous throughout the British Isles. According to this version, Watson named his property 'Keilor', which is Gaelic for 'plenty', after his father's house. Despite the fact that he arrived with 12,000 pounds—a fortune for those days—Watson went bankrupt and returned to Scotland. Keilor became a city in 1961.

Keira, Mount *see* **Mount Keira**

Keith *SA* was named after Keith Stirling, the son of the naturalist, Sir Lancelot Stirling, in 1889.

Kellys Bush *NSW*, is a locality in the Sydney municipality of Hunters Hill. It was originally called Kelly's Paddock. It derives its name from T.H. Kelly who owned about eight hectares of land between Woolwich Road and the Parramatta River in 1892 when a smelting works was first established on the foreshores.

Kellyville *NSW* is a locality in the Baulkham Hills Shire on the north-western outskirts of Sydney. It is named after Hugh Kelly who was one of the pioneers of the locality. Kelly kept a public house in Windsor Road. In 1884 a large subdivision consisting of fifteen grants was surveyed and sold as the Kellyville Estate. The district has been known as Kellyville since that date.

Kelmscott *WA* in Armadale, a town on the outskirts of the Perth metropolitan area, was the name chosen in 1830 by Governor Stirling to honour the birthplace of Archdeacon Scott, who preached the colony's first Christmas Day service at the Rush Church in Perth in 1830.

Kelvin Grove *Q* is a locality in the Brisbane metropolitan area and was named after Kelvin Grove Park in Glasgow by J. Bancroft.

Kembla, Port *see* **Port Kembla**

Kemps Creek *NSW*, a suburb of Liverpool, is named after Anthony Fenn Kemp, an ensign in the New South Wales Corps who arrived in Australia in 1793 and was granted land in 1810.

Kempsey *NSW* was founded by Enoch Rudder, a Sydney merchant, whose interest in the cedar trade first brought him to the Macleay Valley in 1835. He selected the name, Kempsey, because of the valley's similarity to the town of Kempsey in the Severn Valley in Worcestershire, England.

Kenmore *Q* is a locality in the Brisbane metropolitan area and is named after a town in Scotland.

Kensington *NSW*, a locality in the Municipality of Randwick, takes its name from the London borough of that name.

Kensington *SA* is a suburb of the corporate city of Kensington and Norwood in the Adelaide metropolitan area. Kensington was named after Kensington in London. Kensington Palace was the birthplace of Queen Victoria, who succeeded to the throne on the death of her uncle, King William IV, on 20 June 1837 and was crowned in 1838. Kensington was named by Charles Catchlove and surveyed by J.H. Hughes in November, 1838. Catchlove was a part owner of the land with John Brown.

Kensington Gardens and **Kensington Park** *SA* are suburbs of the City of Burnside on the south-eastern outskirts of the Adelaide metropolitan area. They are both named after Kensington in London. For many years Kensington Park was known as Shipster's Paddock after George Frederick Shipster who arrived in September 1843 and bought the land in 1844. Shipsters Road is situated on the western boundary of Kensington Park. Kensington Gardens was formerly known as Pile's Paddock.

Kent Town *SA* is a suburb of the corporate city of Kensington and Norwood in the Adelaide metropolitan area. Kent Town was named after Dr Benjamin Arthur Kent who arrived in South Australia in June 1840 and built 'East Farm Cottage' on this land which he leased from Colonel Torrens (father of R.R. Torrens of the Torrens Title system of land-registration fame). Colonel Torrens was Chairman of the Board of Commissioners appointed by the Crown to manage land sales and emigration to the new province. Kent was in dispute with Colonel Torrens until he finally dropped all claims and bought the land outright in 1854. The land was then subdivided and called Kent Town. Kent returned to England some time in 1859.

Kentish *T*, a district in northern Tasmania, is named after Nathaniel Lipscombe Kentish, a government surveyor who discovered the plains in1842.

Kentucky *NSW* was the name of the pastoral selection made by A.J. Maister in 1840. It is not known if he called it after Kentucky in the United States.

Kenwick *WA* is a suburb of Gosnells, a city in the Perth metropolitan area. It was named after Wallace Bickley's homestead 'Kenwick Park'. Wallace Bickley was the instigator of the formation of a Road Committee in 1868.

Keppel Bay *Q* was named after Admiral Keppel on 25 May 1770, by Captain James Cook.

Kerang *V* is a township beside the Loddon River. Its name comes from an Aboriginal word said to mean 'edible root vegetable'. Other suggested meanings include 'moon', 'cockatoo', and 'parasite'.

Kerrimuir *V* is a suburb of the City of Box Hill in the Melbourne metropolitan area. Kerrimuir is named after a place west of Montrose in Scotland.

Kew *V* is a city about ten kilometres from the centre of Melbourne. Kew's first land sale took place in 1845. No further land sales took place in Kew until 1851, when the discovery of gold sparked off a boom in land. One of the purchasers was N.A. Fenwick, a former goldminer and the founder of Kew. His Kew estate gave the name to the town. He is said to have chosen the name because Kew in England is near Richmond and his estate was near the Melbourne suburb of Richmond. Kew was proclaimed a municipality in1860 and created a borough in 1863. Kew became a city in 1921.

Kewdale *WA* is derived from Kew Street which was one of the first roads through the area. Kewdale was gazetted as a postal district on 7 December 1949.

Keysborough *V* is a suburb of the City of Springvale on the outskirts of the Melbourne metropolitan area. Keysborough is named after George Keys, an Irish immigrant who leased land in the district in the 1840s.

Kiama *NSW* A wide variety of translations of this Aboriginal name has been suggested: 'where the sea makes a noise', 'plenty of food', 'place where many fish may be caught', 'fish may be caught from the rocks' (*kia-ra-mia*), or the name of the Creator Spirit, usually known as Baiame (or by similar names) in New South Wales.

Kiandra *NSW* is an Aboriginal name meaning 'sharp stones used for knives'. It was the site of a gold rush and the diggers became Australia's first skiers.

Kidman Park *SA* is named after Sir Sidney Kidman who owned property in the area.

Kiewa *V*, a township nineteen kilometres south-east of Wodonga, came from Cyanana Warra meaning 'sweet water', which was the Aboriginal name for the Kiewa River at this point.

Kilburn *SA* is a suburb in the City of Enfield in the Adelaide metropolitan area. Kilburn was named after a suburb of London. Portions of it were originally named Chicago after the American city, but it was changed to Kilburn in1930, at the request of residents because of adverse publicity about the American city.

Kilkenny *SA* is a suburb of Woodville, a corporate city in the Adelaide metropolitan area. It was named after the town in Ireland.

Killara *NSW*, a suburb of Ku-ring-gai Municipality, is said to mean 'always there' but there is some dispute as to whether it is an Aboriginal word at all. It could be of Irish derivation.

Kilmore *V*, a township sixty kilometres north of Melbourne, was named in 1841 by a land speculator named William Rutledge. He bought an estate of twenty square kilometres and called it 'Kilmore' after his birthplace in County Cavan, Ireland. A municipality was proclaimed in 1856.

Kimberley District *WA* Following exploration of the far north of W.A. by the expedition led by Alexander Forrest in 1879, it was decided to open up the new district for settlement. The Governor, Sir William Robinson, sent copies of suggested land regulations to the Earl of Kimberley, Secretary of State for the Colonies, and in his despatch said that it was proposed that the new country should be called the Kimberley District. This was approved by the Earl of Kimberley in a telegram dated 26 August 1880.

Kimberley Range *WA* was named by John Forrest on 20 May 1874, in honour of Lord Kimberley, Secretary of State for the Colonies.

Kincumber *NSW* means 'belonging to an old man'.

King Edward River *WA* was named in 1901 by F.S. Brockman, Chief Surveyor of Western Australia in honour of King Edward VII.

King George Sound *WA* was originally named King George III's Sound by George Vancouver, the British explorer who discovered it in 1791.

King Island *T* was named after Governor King of New South Wales by Captain Black who sailed through Bass Strait in 1801.

King, Lake *see* **Lake King**

Kingaroy *Q* From an Aboriginal word meaning 'red ant'. However there is some uncertainty about the origin of the name, some saying that it is a corruption of the Aboriginal word *kinjerroy*, others that it came from an early settler by the name of King. Whatever the actual origin it is generally accepted that the town was named by C.R. Haly, the owner of 'Taabinga' station, in 1846.

Kinglake *V* is a locality in the Shire of Eltham on the outskirts of Melbourne. Kinglake is named after Alexander Kinglake, surveyor and historian, who mapped the way across the mountain top from Queenstown to Glenburn in 1870.

Kings Cross *NSW*, is a locality in the City of Sydney. Kings Cross was Sydney's most prestigious residential district in the first half of last century, when the area was known as Woolloomooloo Hill and contained Sydney's finest mansions. It had become known as Queen's Cross by 1897 and was given its present name in 1905 to honour King Edward VII.

Kings Langley *NSW* is a locality in the City of Blacktown on the western outskirts of Sydney. It was named by Matthew Pearce (1762-1831) who called his land grant after King's Langley Manor House, Hertfordshire, England, where he was born. It was situated on the opposite side of the Windsor Road to the present suburb of Kings Langley.

Kingsbury *V* is a suburb of the City of Preston in the Melbourne metropolitan area. It is named after Private Bruce Kingsbury who won the Victoria Cross in New Guinea where he was killed in 1942.

Kingscote *SA* is the main settlement on Kangaroo Island. It is called after Henry Kingscote, a director of the South Australian Company.

Kingsgrove *NSW* is a suburb of the Sydney municipalities of Canterbury, Rockdale and Hurstville. It was derived from the name 'Kings Grove Farm' given to the grant of 500 acres (200 hectares) to Hannah Laycock by Governor King in August 1804. She named it after Governor King. The grant was a rectangle bounded by Stoney Creek Road, Kingsgrove Road, William Street and Bexley Road. When the railway to East Hills was opened in 1931, one station was named Kingsgrove.

Kingston *ACT* is named in honour of Charles Kingston, one of the founders of the Constitution.

Kingston *Q* was named after an early settler, Charles Kingston, who took up land in 1871.

Kingston *T* was formerly called Browns River. It was officially changed to Kingston at the beginning of 1882, although the latter was being used long before this date. There is an unconfirmed record that Kings Town was applied during the reign of George III, which ended in 1820.

Kingston Park *SA* recalls the Kingston family who lived in 'Marino', Kingston Park, for eighty years. It is now used by the community and has been partly restored by the State Heritage and Conservation Unit. The house was built by Sir George Strickland Kingston, who arrived in South Australia in 1836 as deputy surveyor to Colonel Light. He was the architect of many well-known Adelaide buildings and a member of the Legislative Council. He became Speaker in 1857 and was knighted in 1870. He planted the twin pine trees which still stand on the foreshore of his property and named them after his two sons, Paddy and Charlie.

Kingswood *NSW*, a locality within the boundaries of the City of Penrith, is named after Governor Phillip Gidley King (1758-1808), the third governor of New South Wales. It was formerly known as Crossroads.

Kingswood *SA* is probably named after Kingswood in Gloucestershire, England.

Kirrawee *NSW*, in Sutherland Shire, is an Aboriginal word meaning 'lengthy'. It was known as Bladesville but the name was changed with the opening of the electric railway line in 1939.

Kirribilli *NSW*, a locality in North Sydney, is most likely an anglicisation of an Aboriginal word. It has also been suggested that it was adapted from the name of an early Milson family cottage, 'Carabella'. This could be a happy coincidence.

Kissing Point *NSW*, a locality in the Municipality of Ryde, has long been the subject of romantic tales concerning governors kissing ladies. However, a more likely explanation is that it came from the nautical term 'kissing' which refers to keels touching lightly on the shelving river bed beyond the point. In its early days, the whole district was known as Kissing Point.

Klemzig *SA* is a suburb in the City of Enfield in the Adelaide metropolitan area. Klemzig was named after a town in northern Germany. It was first settled by Lutherans from Silesia under the leadership of Pastor Kavel in 1838. By 1870 most of the original German settlers had moved on to the Barossa Valley. It was changed in 1918 to Gaza, then a World War I battlefield, but reverted in 1935 to Klemzig. The Aboriginal name of the locality was Warkowodli wodli.

Knocker Bay *NT* Entered by P.P. King in April 1818 and probably named by him.

Knox *V* is a residential and industrial city in the foothills of the Dandenong Ranges about twenty-four kilometres from the centre of Melbourne. Knox was named after Brigadier Sir George Knox (1865-1960), a local member of the state parliament from 1927 to 1960, and a councillor from 1923 to 1928 in Fern Tree Gully. Knox became a shire in 1963. It was proclaimed a city in 1969.

Kogarah *NSW* is a municipality on the southern shores of Botany Bay. Kogarah is an Aboriginal term meaning 'place of reeds'.

Kooragang Island *NSW* is located in the Hunter River at Newcastle. The Aboriginal word means 'Aboriginals there' or 'camp there' in the language of the Awabkal or Kattang tribes that lived in the coastal areas from Lake Macquarie to Port Stephens. The Kooragang Estate on Mosquito Island was purchased by the Reverend C. Pleydell N. Wilton. The estate was in existence prior to 1848. The islands of Henhom, Ash, Moscheto, Dempsey, Walsh, Spectacle, Table, Pig and Goat were reclaimed in 1951 by the State Public Works, and amalgamated to form one island, Kooragang Island.

Kooyong *V* is a suburb of the City of Malvern in the Melbourne metropolitan area. 'Kooyong' was the name of a house in Warra Street, Toorak. The word *kooyong* is Aboriginal for 'resting place'.

Koroit *V* is an Aboriginal word for 'fire'. It probably reflects the fact that Australia's most recently active volcano last erupted about 5000 years ago, when local Aboriginals would have seen it.

Korumburra *V*, a township forty-three kilometres south of Warragul, is said to be Aboriginal for 'blowfly'.

Kosciusko, Mount *see* **Mount Kosciusko**

Kununurra *WA* comes from an Aboriginal word meaning 'big waters'. The town was established in the 1960s to serve as the main centre for the Ord River Irrigation Scheme.

Ku-ring-gai *NSW* is a municipality on the upper North Shore of the Sydney metropolitan area. Ku-ring-gai gets its name from the tribe which inhabited the coastal district. Ku-ring-gai Chase was named in 1896. (A chase is an enclosed park). The shire took on the name Ku-ring-gai when it was formed in 1906 and retained the name when it became a municipality in 1928.

Ku-ring-gai, Mount *see* **Mount Ku-ring-gai**

Kurnell *NSW*, in the Sutherland Shire, is thought to be an Aboriginal corruption of the name of John Connell, junior, second landowner but first settler in the shire. John Birnie, the first landowner, did not live on his land.

Kurraba Point *NSW*, a locality in North Sydney, was originally known as Thrupps Point after Lieutenant Alfred Thrupp who received a grant of land there in 1814. Another name for it was Ballast Point because ballast was obtained from the stone cliffs there for ships returning with insufficient cargo to London or those carrying wool which needed extra weight. Stone for Fort Denison was quarried there.

Kurrajong *NSW*, a locality in the Blue Mountains west of Sydney, is named after the well-known tree. The Aboriginals used its strong bark for fishing lines and its roots for food. White pastoralists later found it useful as a shade tree.

Kurralta Park *SA* is a suburb south-west of Adelaide in the corporate city of West Torrens. The subdivision was laid out by Henry Allchurch and sold in 1918. The name was suggested by H.C. Talbot of the Survey Department (now the Department of Lands) after the name Galway was rejected because of duplication. The land was once owned by Dr William Wyatt, Protector of Aborigines and Inspector of Schools under the First Education Act. 'Kurralta' was the name adopted by him for one of his earlier estates and house at Burnside, east of Adelaide. It is an Aboriginal word meaning 'up there'. 'Kurralta House' was designed for Wyatt by Sir George Kingston in 1848.

Kurri Kurri *NSW* means 'very quick' or 'the very first'.

Kwinana *WA* is a town about twenty kilometres south of Fremantle. It is named after SS *Kwinana*, which was wrecked on the shore on 5 May 1922. She caught fire in 1921 and collided with another ship on her way to Fremantle. It was decided that she was not worth repairing and was stripped of her fittings and towed to Garden Island. During a storm she broke her moorings and was blown across Cockburn Sound to the place where her rusting hulk remained a landmark until it was cut down to water level in 1959. It was later filled with limestone and now forms part of a jetty at Kwinana. In 1922, the local postmistress, Clara Wells, first marked mailbags "Kwinana Wreck' at her general store, which is still operating opposite the site of the wreck. The town-site was gazetted in 1937.

Kyabram *V*, a township forty kilometres south-east of Echuca, is derived from an Aboriginal word meaning 'thick forest'.

Kyeemagh *NSW*, a suburb of the municipality of Rockdale, is a very recent name from a polo club which played in this area, called the Kyeema Polo Club. The 'gh' was added to the name at a later date.

Kyneton *V*, a township eighty-five kilometres north-west of Melbourne on the Campaspe River, was named after the village of Kineton near Stratford-on-Avon, England. Gold-diggers in 1851 knew the place as Kyneton. *Kyne* is an archaic word for 'cow'.

Kyogle *NSW* is an Aboriginal name for 'brolga'.

La Perouse *NSW* is named after Jean Francois de Galaup, Comte de la Pérouse, the French navigator who arrived at Botany Bay in January 1788 just as Governor Phillip was leaving it in favour of Sydney Cove. Pére Louis Réceveur, a naturalist and astronomer on La Pérouse's ship, is buried at the place where the French party camped. In 1825, the French erected a monument to La Pérouse, who disappeared after leaving Botany Bay.

Lachlan River *NSW* is named after Governor Lachlan Macquarie. The river was discovered in 1815 by Acting-Surveyor G.W. Evans, and explored by John Oxley in 1814. The Aboriginal name was Callara.

Lady Robinsons Beach *NSW*, a suburb of the Municipality of Rockdale, was named after the wife of Sir Hercules Robinson, Governor of New South Wales, 1872-1879, who used to go horse riding along its sands.

Lake Albacutya *V* is located north of Lake Hindmarsh. The name comes from an Aboriginal word *ngelbakutya* for 'sour quandong', a native plum.

Lake Alexandrina *SA* was named after Princess Alexandrina, later Queen Victoria, by Sturt in 1830. Aboriginal names were Parnka, Kayinga and Mungkuli.

Lake Amadeus *NT* is named after King Amadeus of Spain. Ernest Giles, who discovered the lake in 1872, wished to confer the name of his patron, Ferdinand von Mueller, but Mueller insisted that the lake and nearby mountain be named 'in honour of two enlightened royal patrons of science', King Amadeus and Queen Olga of Spain. Giles described the lake as an 'infernal lake of mud and brine', so perhaps the Baron was fortunate in insisting that it should be named after the royal personage. According to the Australian Broadcasting Corporation, the pronunciation is ah-MAHD-ee-us rather than ah-mah-DAY-us.

Lake Bonney *SA* There are two lakes of this name in South Australia.
  1. Lake Bonney, Riverland, about twenty-four kilometres from Renmark, was discovered and named in 1838 by Joseph Hawdon who, with Charles Bonney, was driving stock overland for the first time from New South Wales to South Australia. The Aboriginal name for this lake was Nookamka. The name was later changed to Barmera (Aboriginal for 'water') because of the confusion with the other Lake Bonney in the south-east of the state near Mount Gambier. But it reverted to Lake Bonney in 1913. The name was changed to 'Lake Bonney, Riverland' in 1982 to avoid confusion.
  2. Lake Bonney, S.E., is located in the South-East of South Australia, west of Mount Gambier and near the coastline. It was named by Governor Gawler during his visit to the area in 1844. Gawler called it after Charles Bonney who had passed near the lake in 1839 while on the first overland journey with stock from Henty's station in what is now Portland in Victoria, to Gumeracha in South Australia. In 1916, consideration was given to changing the name to Coonunda, the name of an adjacent cattle station owned by Peter Begg in the 1850s, but nothing eventuated. The Coonunda National Park now adjoins the lake. The name was changed to Lake Bonney S.E. in 1982 to avoid confusion with Lake Bonney in the Riverland region.

Lake Burley Griffin *ACT* was named after Walter Burley Griffin, a Chicago architect who won a worldwide competition for the design of the federal capital city. The lake is artificial.

Lake Cadibarrawirracanna *SA* is an Aboriginal name and the third longest recognised place name in Australia. According to an authority of the Adelaide Museum, Cadibarrawirracanna means 'stars dancing on the surface of the lake'.

Lake Callabonna *SA* is a salt lake about 560 kilometres north-east of Adelaide. It was originally thought to be part of the 'horseshoe' configuration of Lake Torrens. This was disproved by A.C. Gregory in 1858 when he passed between Lake Callabonna and Lake Blanche. Lake Callabonna was first known as Mulligan, a corruption of the Aboriginal name, Mullachan. The name, Callabonna, was suggested by Dr Stirling in 1894. The lake is the site of world-famous fossil remains of prehistoric animals. Water occasionally fills its southern end and bird colonies, including the banded shrike, breed there.

Lake Cargelligo *NSW* lies 568 kilometres west of Sydney. Cargelligo comes from an Aboriginal word descriptive of the lake, and may even mean 'lake'. When discovered by John Oxley in 1817, he named it Regent's Lake in honour of the Prince Regent. A nearby run owned by Francis Oakes in 1842 was named 'Gagellaga', but when transferred to D. and S. O'Sullivan in 1848 it received its present name, which is given both to the lake and the town. In *The Man Who Sold His Dreaming*, Roland Robinson writes that the Aboriginal, Fred Biggs, of the Ngeamba tribe, told him that the name was the nearest the white man got to saying 'Kartjellakoo' (*kartjell* — coolamon; *akoo* — he had). The name therefore means 'he had a coolamon'.

Lake Corangamite *V*, located between the towns of Colac and Camperdown, gets its name from an Aboriginal word meaning 'bitter water'.

Lake Eacham *Q* A corruption of the Aboriginal name, Yeetcham.

Lake Echo *T* is a descriptive name which was applied when the echo was first heard. At one time it was known as Jones Lake.

Lake Eildon *V*, an artificially created water storage on the Goulburn River, gets its name from the Eildon Hills in Scotland.

Lake Eppalock *V* is an artificially created lake twenty-one kilometres south-east of Bendigo on the Campaspe River.

Lake Eyre *SA* is the largest lake in Australia. It covers nearly 10 000 square kilometres. It is located in the central northern part of South Australia, which generally receives less than 120 millimetres of rain a year. As a result, its surface is usually dry and covered with a salt crust up to five metres thick. The lake is filled only in times of unusual rainfall. At these times the lake attracts bird life such as pelicans and even — amazing as it may seem — fish. The explorer, Edward John Eyre, was the first European to find the lake which he saw for the first time on 14 August 1840 from a high bank now called Eyre Lookout. It was later named in his honour by Governor MacDonnell. The Aboriginal name was Kati-tanda. The dry salt lake was used by Donald Campbell to break a land speed record in 1960.

Lake Frome *SA* is one of the chain of salt lakes in the north-eastern corner of South Australia. The lake is a dry salt flat for most of the year. It measures about fifty by one hundred kilometres. The explorer, Edward John Eyre, was the first European to sight the lake in 1840. It is named after E.C. Frome, Surveyor-General of South Australia. Frome abandoned an expedition to Central Australia in 1843 when he found Lake Frome blocking his way.

Lake George *NSW* was discovered in 1820 by a pioneer settler, Joseph Wild. Governor Macquarie visited it later in 1820 and named it after King George IV. Since 1820 it has dried up completely at least four times.

Lake Hindmarsh *V*, on the Wimmera River near Jeparit, was discovered in 1838 by Edward John Eyre, who named it after the Governor of South Australia.

Lake Hume *V, NSW*, an artificially created lake on the Murray River east of Wodonga, is named after the explorer, Hamilton Hume.

Lake Illawarra *NSW* was originally named Tom Thumb's Lagoon by George Bass, Matthew Flinders and William Martin who landed near the lake in their tiny three metre craft, the *Tom Thumb*, in March 1796 while in search of a reported river. It was also known for a while as Big Tom Lagoon. The name, Lake Illawarra, was finally adopted and the name Tom Thumb's Lagoon was given to a stretch of water at Port Kembla. Illawarra is an adaptation of an Aboriginal word which can be translated as 'camp by the sea' although other meanings can also be given.

Lake King *V* in Gippsland was named by Strzelecki in 1840 after Phillip Park King, who became Australia's first rear-admiral.

Lake Lefroy *WA*, a salt lake, was discovered by Charles Hunt in 1864 and named after Henry Maxwell Lefroy, who had been engaged in exploratory work in that region a year earlier. Lefroy was assistant superintendent of the Fremantle gaol.

Lake Macquarie *NSW* is named after Governor Macquarie. The Aboriginal name was Awaba, meaning 'level surface'.

Lake Mungo *NSW* gets its name from *mungo,* an Aboriginal word for 'tall rushes'.

Lake Pedder *T* was named on 11 March 1835 by the surveyor, John Helder Wedge, after Sir John Lewes Pedder, a chief justice of Tasmania.

Lake Phillipson *SA* is named after N.E. Phillipson who owned the 'Beltana' station. It was discovered and named by J. Ross on 24 June 1874.

Lake St Clair *T* was named by Surveyor-General, George Frankland, in 1835 after the St Clair family of Loch Lomond, Scotland. The Aboriginal name was Leeawulena.

Lake Torrens *SA* was discovered by E.J. Eyre in 1839 and later named by him after Colonel Robert Torrens, Chairman of the South Australia Colonization Commission, and father of Sir Robert Richard Torrens who became Premier of South Australia in 1857. Lake Torrens was originally known as Horseshoe Lake. It was thought that what are now Lake Torrens, Lake Eyre, Lake Gregory, Lake Blanche and Lake Callabonna were all one lake.

Lake Tyrrell *V*, north of the township of Sea Lake, was discovered in 1838 by E.J. Eyre, who named it after a squatter in the Port Phillip district.

Lake Victoria *V* in Gippsland was discovered in 1840 by Angus McMillan, who named it after Queen Victoria.

Lake Wellington *V*, east of Sale in Gippsland, was discovered in 1840 by Angus McMillan, who named it after the Duke of Wellington.

Lake Wendouree *V* is a small lake at Ballarat, which was used as the rowing venue for the 1956 Melbourne Olympic Games. It was originally called Yuille's Swamp after a squatter named William Yuille, who built a hut on the banks of the Black Swamp in 1838. He believed the Aboriginals called the place Wendouree. In fact, they were saying *wendaaree,* meaning 'go away'.

Lakemba *NSW* is a suburb of the Sydney municipality of Canterbury. It was named after the house of an early settler, Benjamin Taylor, which stood in Haldon Street where the

Sussan store is today. He called his estate 'Lakemba' after an island in the Pacific Ocean where his father was a missionary.

Lakes Entrance *V*, a fishing and resort centre, takes its name from the entrance to Gippsland Lakes from Bass Strait, which John Reeves discovered in 1841. It was known as Cunninghame until 1889.

Lalor *V* is a locality in Whittlesea Shire on the outskirts of Melbourne. Lalor perpetuates the name of Peter Lalor who led the rebels at the Eureka Stockade in 1854 and later became a member of parliament.

Lalor Park *NSW* is a locality in the City of Blacktown on the western outskirts of Sydney. It is named after the Lalor family, who owned property in the area. George A. Lalor was Blacktown Shire President from 1914-1928 and 1920-1937.

Lamington Plateau *Q* was named after Lord Lamington, Governor of Queensland from 1895-1901, the man after whom lamington cakes are also named.

Lane Cove *NSW* is a municipality on the North Shore of Sydney Harbour. Lane Cove seems to have been christened very early in Sydney's history. The first mention of it is in William Bradley's journal of 2 February 1788 during the course of Captain Hunter's survey of the upper harbour. Unfortunately Bradley neglected to say how the cove came to be named. It is possible that it was so named by Governor Phillip in the early days of the colony in honour of his great friend, John Lane, a well-to-do London merchant. However, there is no proof whatsoever as to the correct origin of the name Lane Cove.

Langford *WA*, a locality in Gosnells, a city in the Perth metropolitan area, honours the Langford family. W.H. Langford served as a councillor for over forty-three years before his retirement in 1976.

Lansdowne *NSW* is a suburb of Fairfield and Bankstown cities on the south-western outskirts of the Sydney metropolitan area. Lansdown is named after the historic bridge which crosses Prospect Creek. The name was bestowed by Sir Richard Bourke (Governor of New South Wales) when the bridge was opened in 1836 and called after Lord Lansdowne.

Lansvale *NSW* is a suburb of Fairfield on the south-western outskirts of the Sydney metropolitan area. It was the name chosen by local residents when the area was made a postal district. Local residents were given a choice of a district name from a list of suggestions. Lansvale is an amalgam of Lansdowne and Canley Vale.

Lapstone *NSW*, a locality in the Blue Mountains west of Sydney, got its name because the early road makers noticed the similarity of the rocks in the locality to the lap stones used by the cobblers of that era and the name Lapstone Hill was given to the steep incline in the foothills of the Blue Mountains. In the late 1950s, a large part of the escarpment east of Glenbrook was opened by the late Arthur Hand for residential development and the Blue Mountains City Council formally named it as the Town of Lapstone in 1964.

Lara *V* is a residential district near Lara Lake north of Geelong. It was also known as Duck Ponds and Hovells Creek. The Aboriginal word, *larra*, means 'stone' or 'building of stones'.

Largs Bay *SA* is named after Largs on the Firth of Clyde, Scotland.

Latham *ACT* honours Sir John Greig Latham (1877-1964), Chief Justice from 1935 to 1952.

Latrobe *T* was named in 1861 after Charles La Trobe, a lieutenant-governor of Victoria.

Latrobe Valley *V* is named after Charles Joseph La Trobe, who served as administrator for a short period in 1846 and 1847, and later became Lieutenant-Governor of Victoria. His name is spelt as two words, but the valley as only one.

Latta, Port *see* **Port Latta**

Launceston *T* was named in honour of the birthplace of Governor King of New South Wales. The decision was made in Sydney some time in 1807. The local pronunciation is LON-sest-in rather than LAWN-sest-in.

Lavender Bay *NSW*, a locality in North Sydney, is named after George Lavender who married Susannah Blue, the daughter of Billy Blue. Like his father-in-law, George Lavender also became a boatman, ferrying passengers across the harbour and back again. After selling his cottage at Lavender Bay, he went to live at the Commodore Hotel in Blues Point Road where he died in 1851, at the age of sixty-six. He and his wife are both buried at St Thomas's Park in West Street, along with other North Sydney pioneers.

Laverton *WA* is named after Dr Laver, a local pioneer.

Lawley, Mount *see* **Mount Lawley**

Lawson *NSW*, a locality in the Blue Mountains west of Sydney, was named after William Lawson (1774-1850), one of the three explorers who were the first white men to successfully cross the Blue Mountains in 1813.

Learmonth *WA* is named after Charles Learmonth, a pilot in World War II, who was killed when his Beaufort plane crashed. He was able to radio the events leading up to the crash, thus explaining several earlier accidents and helping to eliminate the cause.

Leeming *WA*, a locality in the City of Melville in the Perth metropolitan area, commemorates George Leeming, a surveyor who laid out the roads for the Jandakot Agricultural Area in 1889. The name was adopted in1971.

Leeton *NSW* is named after Charles Alfred Lee, the Minister for Works in the New South Wales Government, who headed the committee that planned the Murrumbidgee Irrigation Area. Leeton was designed by the American architect Walter Burley Griffin, the designer of Canberra. It dates from 1912, the year the Murrumbidgee Irrigation Area scheme was officially opened.

Leeuwin, Cape *see* **Cape Leeuwin**

Lefroy, Lake *see* **Lake Lefroy**

Leichhardt *NSW* is an inner city municipality in the Sydney metropolitan area. Leichhardt was named by Walter Beams who bought land in the area and called it after his old friend, the ill-fated, German-born explorer Ludwig Leichhardt (1813-1848) who disappeared without trace while on a journey across Australia. The area had previously been known as Piperston because of the grant made to Captain Piper and Ensign Hugh Piper in 1811.

Leigh Creek *SA* is named after Harry Leigh, who was the head stockman for Alexander Glen, who took up a cattle run in this part of the country in 1856. The town of Copley, named after the Honourable W. Copley M.L.C., Commissioner of Crown Lands 1891, was proclaimed on 27 August 1891. The location of Copley was previously known unofficially as Leigh Creek. The original town at the coalmining area known as Leigh Creek was moved south so the mining operations could be extended. The new town was proclaimed on 5 October 1979 as Leigh Creek South.

Lemon Tree Passage *NSW* was so named because lemon trees (the origin of which remains a mystery) were found by early settlers to be growing on the point.

Lenah Valley *T* was formerly known as Kangaroo Valley, but probably because of the duplication problem, the present name was given about 1922, it being the local Aboriginal name for 'kangaroo'.

Lenswood Valley *SA* was originally part of Forest Range. It was decided to split the settlements at Forest Range into two and call them Forest Range and Maralinga. Maralinga was unacceptable as a name because a place by that name already existed. At this time (1915) a battle had taken place near Lens, Pas-de-Calais, France. The name Lenswood was derived from this incident and became official in 1917.

Leongatha *V*, a township in South Gippsland, is believed to have been derived from an Aboriginal word for 'teeth'.

Leopold *V* is a farming district ten kilometres south-east of Geelong. It was called Kensington until it was renamed in 1892 after Queen Victoria's son, the Duke of Albany.

Leschenault Estuary *WA* was discovered by De Freycinet in 1803 and named after the botanist on the *Geographe*.

Lesmurdie *WA* was named by Archibald Sanderson, a Perth journalist who bought land there in 1897. He named his property 'Lesmurdie' after a shooting-box in Banffshire Scotland which his father rented for shooting. Sanderson constructed an imposing mansion on his land, and named it 'Lesmurdie House'. In 1927 he and his wife were host and hostess to the Duke and Duchess of York, and this occasion must certainly have helped to put the name on the map.

Lethbridge Park *NSW* is a locality in the City of Blacktown on the western outskirts of Sydney. It is named after the Lethbridge family. Robert Copeland Lethbridge settled at Werrington on the land grant made on 1 January 1806 by Governor King to his daughter Mary who was Lethbridge's wife.

Leumeah *NSW* is an Aboriginal word meaning 'here I rest'.

Leura *NSW*, a locality in the Blue Mountains west of Sydney, was originally to be named Lurline after the daughter of the subdivider of the area, but when the railway station opened in 1891 it was called Leura.

Leven River *T* was probably named about 1828 by Hellyer, a surveyor with the Van Diemens Land Company, after the Scottish River.

Lewisham *NSW*, a suburb of Marrickville, is believed to have been named by Judge Joshua Josephson who owned land in that area. Lewisham, once in Kent, is now a South-East London suburb.

Lidcombe *NSW*, a suburb of Auburn, has changed its name more frequently than any place in New South Wales. When the railway station was opened in 1859 it was called Haslam's Creek. It was renamed Rookwood in 1878. Local people objected to this name because of its association with Rookwood Cemetery. After much debate and many suggestions the name Lidcombe was finally adopted in 1913. It is a combination of the names of Mayor Lidbury and a former mayor called Larcombe.

Liddell *NSW* is probably named after Liddel Waters which joins the River Esk about eighteen kilometres north of Carlisle, England.

Lightning Ridge *NSW* is said to have got its name from an incident in which a flock of sheep in the area was struck by lightning. The opal fields were first opened up between 1905 and 1915.

Lilli Pilli *NSW* in the Sutherland Shire is named after the Myrtle Tree (Eugenia smithii).

Lilydale *V* is a township by Olinda Creek, thirty-eight kilometres east of Melbourne. This was one of the earliest wine-growing areas in Victoria. It was originally named Yering after William Ryrie's 'Yering' cattle run. The vines were planted at 'Yering' in 1838. According to one story, the town that arose was supposedly called Lily, after the wife of Paul de Castella, who has been described as the first and most distinguished viticulturist. According to another version, John Hardy, a government surveyor, was surveying the site for a town in 1859 when Mrs Hodgkinson, wife of his departmental superior, visited the site. She suggested the name, Lilly Dale, after a popular song the chainmen had been singing and because there were lilies growing in the pools of the creek. After she had gone, Hardy wrote the name down as Lillydale, with a double 'l', which has caused confusion ever since. The shire is still spelt 'Lillydale' and the town is spelt 'Lilydale'.

Lincoln, Port *see* **Port Lincoln**

Linden *NSW*, a locality in the Blue Mountains west of Sydney, was taken from the German word for lime-trees. It was said to have been named in 1873.

Lindfield *NSW*, a suburb of Ku-ring-gai Municipality, gets its name from a house near the station, between Bent and Balfour Streets. It was named Lindfield after the native town of Mr List, the owner. Mr William Cowan, first president of the Shire, afterwards lived there.

Lindisfarne *T* was named after 'Lindisfarne House', a property adjoining Rosny in the 1820s. This in turn was named after the island off Northumbria.

Lismore *NSW* got its name from a pastoralist, William Wilson, who was granted land in the area in 1843. He named his station after the Isle of Lismore off the coast of Scotland. The name in Gaelic means 'great garden'. Lismore became a municipality in 1879 and a city in 1946.

Lithgow *NSW* is named after William Lithgow (1784-1864), an auditor-general of New South Wales.

Little Beach *NSW* Bark taken from trees standing opposite the beach was once used by Aboriginals to make canoes. Halifax Park, named in memory of the WWII aeroplane of the same name and situated in the vicinity, was established by Geoffrey Wikner in 1947.

Little Desert *V* is a national park south-west of Dimboola. The average rainfall is about 400 millimetres a year. This is higher than true arid areas, so the park is not really desert. The name came from early settlers, who found its soil unsuitable for crops. Since then, its special value has been recognised and it is no longer regarded as useless scrub. Heath, broombush and mallee eucalypts grow profusely, and more than 200 species of birds, as well as possums, feather-tailed gliders and other animals, have been recorded.

Liverpool *NSW* is a city on the outskirts of the Sydney metropolitan area. Liverpool was named by Governor Lachlan Macquarie in 1810. He named it in honour of the Earl of Liverpool who was Secretary for the Colonies.

Lobethal *SA* was named by its original Prussian settlers in 1842. *Lobe* means 'praise' and *thal* means 'valley'. So Lobethal means 'valley of praise'. During WWI the name, Lobethal, was changed to Tweedvale (tweed from the woollen mills being the chief industry, and 'vale' meaning 'valley'), but it was changed back to Lobethal in December 1935.

Lockyer Valley *Q* is named after Edmund Lockyer who was dispatched by Governor Brisbane to explore the Brisbane River in 1825. Lockyer investigated a reported sighting near Fernvale Bridge of a tribe of white men with bows and arrows. He did not find a tribe of white men, but he explored the foothills of Mount Brisbane and discovered Lockyer Creek. His findings upset Oxley's theory about the Brisbane river draining into an inland sea.

Loftus *NSW*, in the Sutherland Shire, is named after Lord Augustus Loftus, Governor of New South Wales, 1879-1885.

Lofty, Mount *see* **Mount Lofty**

Logan, Loganholme and Loganlea *Q* are all located in the Brisbane metropolitan area and named after Captain Patrick Logan (1791-1830), the commandant of the Moreton Bay convict settlement from 1825 until he was killed in 1830. Logan also explored much of the Brisbane area. Logan City is Queensland's third largest city and straddles the Pacific Highway eighteen kilometres south of Brisbane. The city took its name initially from the Logan River which formed its southern boundary for thirty-two kilometres. The river, in turn, is named after its discoverer, Commandant Patrick Logan. When Logan discovered the river on 21 August 1826, he named it the Darling after Governor Darling, but the Governor changed it in recognition of Logan's discovery. Logan is remembered both for his journeys of exploration around the Brisbane area and for his brutality as a commander of a difficult convict penal settlement. Logan died in mysterious circumstances. He may have been killed by his own convicts, but it is more probable that he was killed by Aboriginals.

Long Bay *NSW*, a locality in the Municipality of Randwick, was named by Governor Phillip. The prison there opened in 1900 although construction commenced in 1898.

Long Reef *NSW* is a locality in Warringah Shire which gets its name from the long, rocky reef which juts out into the sea.

Longbottom *NSW* is a locality in the Sydney municipality of Concord. It derived its name from the Longbottom stockade set up by Governor Phillip as a guard post on the way to Parramatta. Bottom is an old English term for low-lying swampy land. The name was later applied to 'Longbottom Farm' established by the Government in the Concord-Burwood area.

Longford *T* was probably named by Roderic O'Conjor, a government surveyor who received substantial grants of land in the area. It seems likely that the country around the junction of the Lake and South Esk rivers reminded him of the area around the junction of two rivers at Longford, the O'Connor home district in Ireland. The district was first known as Norfolk Plains, and was later called LaTour after a leading member of the Cressy Company which carried on large scale farming from 1826 to 1856. Longford was officially named by Governor Arthur in May 1830.

Longreach *Q* was gazetted in 1887. It was named after a 'long reach' of the nearby Thomson River.

Looking Glass Point *NSW*, a locality in the Municipality of Ryde, recalls an incident that took place on 15 February

1788, when Governor Phillip gave a mirror to an armed Aboriginal who came upon the party of explorers while they were having breakfast. The man was intrigued with the gift and looked behind the glass to see if anybody was there.

Lord Howe Island was visited by Lieutenant Henry Lidgbird Ball, Commander of the *Supply*, in 1788 on his way to Norfolk Island from Port Jackson. He named it in honour of Lord Howe, a British admiral. The first settlers landed in 1834.

Lorne *V* was first named Loutit Bay after Captain Loutit who called here while on a voyage to London in 1841 with the first consignment of wool from the district. When the township became a seaside resort in 1871 the name was changed to Lorne, after the town in Argyllshire, Scotland.

Lovely Banks *V* is a rural locality north of Geelong. It is a descriptive name but may also recall Lovely Banks Inn which stood by the Jordan River in Van Diemen's Land in the early nineteenth century.

Loxton *SA* was first known as Loxton's Hut, from a hut built by W.C. Loxton, a boundary rider on the 'Bookpunong' station. The town was not proclaimed until 1907 and was named Loxton by Governor Le Hunte.

Lucas Heights *NSW*, in the Sutherland Shire, is named after John Lucas, a flourmiller at Liverpool who was granted land at 'the head of unnamed stream falling into George's River' in 1823. He built a water-driven mill for grinding corn from the Illawarra farms. Small ships brought the corn up the coast into Botany Bay, Georges River and Woronora River.

Lucinda; Lucinda Point *Q* The name of the point and the settlement came from that of a government steam yacht.

Lugarno *NSW*, a suburb of Hurstville, was probably named after Lake Lugano (note the different spelling) on the border of Italy and Switzerland, because of a resemblance to this lake. It was probably named by Thomas Holt or his employee, James Murphy, the manager of Holt's Sutherland estate in the 1870s. The name applied to the general area of that reach of water. The name was definitely applied to the northern shore, its present location, in the early 1900s due to the presence of a ferry service and boatshed. In the 1870s, Thomas Holt was promoting the Georges River Water Supply Scheme, whereby Georges River was to be dammed to provide the water supply for Sydney. Thomas Holt had visited Europe from 1866 to 1868 and he said he had seen the most beautiful lakes in England, Italy and Switzerland, but they could not be as beautiful as the series of lakes formed if a dam was built across the Georges River at Sylvania; so it is quite likely that the names of Como and Lugarno were given at that time. The high cliffs bordering the river with its many windings gave the appearance of a series of lakes.

Luina *T*, now a tin-mining town, was first listed in the 1914 edition of *Walch's Almanac*, having previously been known as Whyte River, the name of the stream on which it is situated. It was not proclaimed as a town until 9 June 1971. *Luina* is an Aboriginal word meaning 'blue-head wren'.

Lutwyche *Q* is a locality in the Brisbane metropolitan area and is named in honour of Mr Justice Lutwyche, the resident judge who swore in Sir George Ferguson Bowen as the first governor of Queensland in 1859.

Lyell, Mount *see* **Mount Lyell**

Lyndoch and Lyndoch Valley *SA* are located in the Barossa Valley. It was named in 1837 by Colonel Light who called it after his friend General Thomas Graham, later Lord Lynedoch (correct spelling). Light was serving as aide-de-camp to the Duke of Wellington in the Peninsular War when he met Graham in 1811. Both men took part in the victorious campaign against the French at the village of Barossa, south-east of Cadiz. The first recorded reference to the township is spelt 'Lyndock'.

Lyneham *ACT* is named after Sir William Lyne, one of the founders of the constitution.

Lyons *ACT* honours Joseph Aloysius Lyons (1879-1939), a former Tasmanian premier who served as Prime Minister of Australia from 1931 to 1939.

Macarthur *ACT* is called after John Macarthur (1767-1834), one of the main founders of the merino wool industry in Australia.

McArthur River *NT* was named by Leichhardt, on 21 September 1845 while on his first expedition, after James and William Macarthur of Camden who were good supporters. James Macarthur was the fourth son of Captain John Macarthur and William the fifth son. The name applied to the river was correctly spelt by Leichhardt, but has since been changed. It is spelt McArthur on Northern Territory plans but it is sometimes spelt Macarthur by local people.

McCallums Hill *NSW* is a suburb of the Sydney municipality of Canterbury. It is located on land along Moorefields Road, extending from Bower Street to Stoddard Street which was owned last century by two sisters, Kezia and Julia Henderson, who married the McCallum brothers, John and Malcolm, merchants of Melbourne. They formed a trust for their children and when it was finally broken up in 1920 this land was sold as the McCallum's Trust. The suburb still carries their name.

Macclesfield *SA* was named after the Earl of Macclesfield.

Macdonaldtown *NSW*, a Sydney suburb, was named after an ironmonger whose home was near the site of the present railway station.

Macdonnell, Port *see* **Port Macdonnell**

MacDonnell Ranges *NT* were discovered in 1860 by John McDouall Stuart, who named them after Sir Richard MacDonnell, a South Australian governor.

Macedon, Mount *see* **Mount Macedon**

Macgregor *ACT* is called after Sir William Macgregor (1846-1919) who was Governor of Queensland from 1909 to 1914.

Macintyre River *NSW* was named by Allan Cunningham in 1827 after Peter Macintyre, manager of the 'Segenhoe' station on the upper Hunter River. The Aboriginal name was Karaula.

Mackay *Q* is a city about 100 kilometres north of Brisbane. In 1860 Captain John Mackay led a party from Armidale on an overland expedition in search of pastoral country. Mackay's party reached the summit of the main coastal range on 16 May and looked out over a rich tropical valley. On 26 February 1862 Mackay returned to the area with 1200 head of cattle to establish the initial settlement of Greenmount. This settlement was about twenty kilometres from the mouth of the river, which was originally called the Mackay, but was later renamed the Pioneer in honour of a visit by HMS *Pioneer* under the command of Commodore Burnett. It was not until after Mackay's departure that the township on the banks of the river was surveyed and given his name. Finally, a word on pronunciation. North Queenslanders pronounce it May-KEYE (rhymes with 'eye') rather than May-KAY.

McKellar *ACT* is named after Gerald Colin McKellar (1903-1970), a leading member of the Senate from 1958 to 1970.

MacKenzies Bay *NSW* at Tamarama recalls Alexander Kenneth MacKenzie who arrived in Sydney with his family in 1822. His grandson, named after him, established MacKenzie's Dairy on land near Denham Street, Bondi. By 1904 it had the largest number of milking machines of any dairy in the world. In later years when paddocks outside the dairy were sold, cows were driven to the 'boot' of Tamarama which is known today as MacKenzies Bay.

Macksville *NSW* was notified as a private town in 1908. It was supposed to have been named after the four owners of the land subdivided to form the town. Their names were MacGuinnes, McKay and McKenna.

McLaren Vale *SA* may have been named after John McLaren, a member of the survey department who divided the south into three districts and opened them to the public in February 1839. However, it may have been named after David McLaren, colonial manager of the South Australian Company from 1837 to 1840. The former version is more likely correct because it is doubtful if David McLaren, in the company of others, ventured further south than Happy Valley in 1837. Aboriginals knew McLaren Vale as Myallinna Dooronga according to Rodney Cockburn's *What's In a Name*, but H.C. Talbot says the Aboriginal name was Tattachilla.

Maclean *NSW* Named after the surveyor who planned the first settlement. Before this it was known as Rocky Mouth.

Macleay, Point; Macleay Range; Macleay River *NSW* After Alexander Macleay, scientist and official, who played an important part in the early history of the colony. The Macleay Islands in Moreton Bay, Queensland, were also named after him. The Forest Kingfisher, *Halcyon macleayi*, was also named after him. The Aboriginal name of Macleay Point was Yarrandabby.

McMahons Point *NSW*, a locality in North Sydney, is named after a member of the first North Sydney Council which was incorporated in 1890. Michael McMahon was a brush and comb manufacturer and land speculator who moved to the point in 1871.

McPherson Range *Q, NSW*. Named in 1828 by Allan Cunningham after Major Duncan McPherson of the 39th Regiment.

Macquarie *ACT* recalls Lachlan Macquarie (1762-1824) who was Governor of New South Wales from 1810 to 1921.

Macquarie Harbour *T* was named after Governor Macquarie by Captain James Kelly on his voyage around Van Diemen's Land in an open whaleboat in 1815.

Macquarie Island is an Australian Territory about 1450 kilometres south-east of Tasmania and 1400 kilometres from the Antarctic continent. It was discovered in 1810 by Frederick Hasselburg, an Australian sealer, who named it after Lachlan Macquarie.

Macquarie, Lake *see* **Lake Macquarie**

Macquarie, Port *see* **Port Macquarie**

McRae *V* is called after Andrew McRae, who took up land in 1843 at Arthurs Seat near the peninsula seaside suburb that now bears his name. His homestead, built in 1844-45, is now classified as part of the National Estate.

Maddington *WA* in Gosnells, a city in the Perth metropolitan area, was named by John Randall Phillips, who called his property here 'Maddington Farm' (presumably after the outer London suburb of that name) as early as 1832. 'Maddington Park' homestead, a gracious two-storey house which Major W. Nairn started to build before 1835 still stands in Burslem Drive. It was completed by Captain Roe, the state Surveyor-General at that time, and leased to the Harris Family. It is now listed as part of the National Heritage.

Mafeking *V* is an old goldfield on the south-east spur of Mount William where gold was first discovered in 1900. It is named after the town in South Africa besieged by the Boers. Baden-Powell commanded the defenders.

Maffra *V* is a township seventeen kilometres north-west of Sale on the Macalister River. One of the early settlers, William Bradley, had served in the Peninsular War and recollected a pleasant stay at the town of Mafra in Portugal. He gave the name to his property. When the town-site was surveyed in 1865, Bradley's choice of name was adopted, but suffered in spelling.

Magill *SA* is a suburb within both the corporate cities of Burnside and Campbelltown in the Adelaide metropolitan area. Magill was the first of the Burnside foothill villages to be subdivided. The Makgill Estate, as it was then called, was owned jointly by Robert Cock and William Ferguson, two Scots immigrants who met on board the *Buffalo* and pooled their resources in several investments, including the Magill project. They named the farm after Mrs Cock's trustee, David Maitland Makgill of Fifeshire. A plan headed

'Makgill' was deposited in the General Registry Office in 1855. With no explanation, the 'k' has since been dropped.

Magnetic Island *Q* is the largest northern resort island on the Great Barrier Reef. Captain Cook named it in 1770 because he wrongly believed the island was affecting his compass.

Mahogany Creek *WA* in Mundaring Shire near Perth received its name when explorers called the jarrah trees mahogany because of the similarity of the two woods. An inn was built here in 1833, located in a small valley surrounded by 'mahogany trees'. The name, Mahogany Creek, was first recorded by George Smythe, while surveying for the York Road in 1835. The locality was named after the creek.

Maitland *NSW* is named after Frederick Maitland, Earl of Lauderdale. There have been changes and some confusion in the names given to this city. The first settlement was formed by 'eleven well-behaved convicts' between 1818 and 1821. The part eventually known as West Maitland was known first as Shancks Forest Plains, later The Camp, then as Molly Morgan's Plains. Mary or Molly Jones married William Morgan and was transported to New South Wales in 1790 (on the Second Fleet) for stealing 'thirty-six clippings of hempen yarn'. She escaped and entered into a bigamous marriage in England. When transported a second time in 1803 (for unpaid debts) she settled near Parramatta. In 1818 she was given a government land grant upon which the West Maitland settlement was founded. Molly Morgan Plains became the local name for Wallis Plains, named in 1818 after Captain James Wallis, the Commandant at Newcastle. Molly Morgan married a third time in 1822 and died in 1835. The Government town (now East Maitland) was surveyed by George White in 1829 and named Maitland, while the early settlement remained as Wallis Plains. But as Wallis Plains was also known as Maitland some confusion arose and in 1835 the names East Maitland and West Maitland were officially adopted. The Aboriginal name for Wallis Plains was Cooloogoolooheit while the land around Morpeth was called Illaluang.

Majors Bay *NSW* is a locality in the Sydney municipality of Concord. It was named in honour of Major George Johnston, father-in-law of Isaac Nichols, an early settler.

Malabar *NSW*, a locality in the Municipality of Randwick, is named after a Burns Philp motor vessel which was stranded on the north side of Long Bay on 2 April 1931.

Maldon *V* is a township thirty-two kilometres south-west of Bendigo in an old goldmining area of historic significance.

It was first called by its Aboriginal name Tarrangawer, which means 'big mountain'. After the discovery of gold in December 1853, it was named in 1856 after a port in Essex, meaning 'hill surmounted by a monument or cross'.

Mallee *V*, a large district in north-west Victoria, gets its name from a type of eucalypt common in the district.

Malvern *V* is a city about ten kilometres south-east of the centre of Melbourne. The first settler was John Gardiner, who brought 400 cattle overland from New South Wales in 1835, and set up a station. The district was originally called Gardiner, but Malvern replaced Gardiner after a village was established. The name 'Malvern' was chosen because the area at that time resembled the Malvern Hills in England. Local government in the area first began in 1856 with the creation of the Gardiner District Roads Board; this became the Shire of Gardiner in 1871. The name changed to the Shire of Malvern in 1878. Malvern became a town in 1901, and a city in 1911.

Mandurah *WA* comes from an Aboriginal name, Mandjar, meaning 'trading place'. It was one of the great sites where bartering took place between the tribes; tools, ornaments, weapons, shells, and other objects being exchanged in a trading system that had links across the continent. The name for the Peel Inlet here was taken from an early settler, Thomas Peel, who adopted the Aboriginal name for his property, 'Mandurah House'.

Manifold Heights *V* is part of the City of Geelong West. Thomas and Peter Manifold first squatted on this land with sheep brought ashore in July 1836 at Point Henry.

Manly *NSW* is a municipality on the northern shore of Sydney Harbour. It has one of Sydney's best known surf beaches. Manly can trace its name back to Governor Phillip who bestowed the name Manly Cove because he was impressed by the manly behaviour of the Aboriginals in the area. The name was frequently used in early accounts of the settlement. When, in the 1850s, Henry Gilbert Smith began to develop a tourist resort he tried to introduce the name Brighton, but the older name for this part of the harbour persisted. Manly was proclaimed a municipality in 1877, though in 1885 the term Village of Manly was adopted.

The Corso, Manly's oddly named main street, is said to derive from the Italian word *corso* which designates a place to race horses or the principal street of a town where horse-drawn carriages used to parade and festivals were held in the larger towns of Italy. The best known of these is the Corso in Rome.

Manning River *NSW* was named by Robert Dawson, the first agent of the Australian Agricultural Company in honour of the Governor of the company, Sir Herbert Manning.

Manns Point *NSW*, a locality in the Municipality of Lane Cove, is a pleasant picnic spot named after the Mann family. Mann obtained the signatures of Greenwich electors for the 1884 petition seeking local self-government for Lane Cove.

Mannum *SA* is an Aboriginal name which is possibly a contraction of Manumph, the meaning of which is not known. According to *Mannum Yesterday* by Bevan and Vaughan, Manyump was the name applied to an Aboriginal camping ground in the vicinity.

Mansfield *V* is a rural township by Fords Creek north-east of Eildon Reservoir. Edward Eyre Mansfield lived there in 1841. The name occurs in Nottinghamshire, England, and Ayreshire, Scotland.

Mansfield Park *SA* is a suburb in the corporate city of Enfield in the Adelaide metropolitan area. Mansfield Park was probably derived from Mansfield in Ayreshire, Scotland. Mansfield Park was subdivided in 1923.

Maralinga *SA* was the site of seven British-Australian nuclear test explosions in 1956 and 1957. Maralinga is eighty kilometres north-east of Ooldea, in south-western South Australia. It lies 1200 kilometres north-west of Adelaide, and 800 kilometres west of Woomera. The name is derived from an Aboriginal term meaning 'thunder'. The village built at Maralinga during the tests has been abandoned and is now a ghost town. The *Maralinga Tjarutja Land Rights Act* (No. 3 of 1984) brought about the vesting and granting of the fee simple in the land to the Aboriginals, who were the traditional owners.

Marayong *NSW* is a locality in the City of Blacktown on the western outskirts of Sydney. It is the Aboriginal name for 'emu'. The name was adopted by the Railways Department in 1922.

Marble Bar *WA* derives its name from a nearby rocky outcrop of jasper/quartzite on the nearby Coogan River, erroneously named 'marble bar'. Marble Bar was known as the hottest place in Australia because for 160 days from 31 October 1923 to 7 April 1924 the temperature did not fall below 38°C (100°F). The bar was thought to have been named by Surveyor C. Crossland in 1908.

Mareeba *Q* is an Aboriginal name meaning 'meeting of the waters'.

Maria Island *T* was named in 1642 by the Dutch explorer Abel Tasman in honour of the wife of Anthony Van Diemen, the Governor-in-Chief of the Dutch East India Company in Batavia who sent Tasman on his expedition.

Maribyrnong River *V* was surveyed by Charles Grimes in 1803. The Aboriginal name, Mirring Gnay Birnong, means 'saltwater river'.

Marion *SA* is a corporate city in the south-western part of the Adelaide metropolitan area. It is located about ten kilometres from the centre of Adelaide. Marion is of uncertain derivation. Although there is no concrete evidence for the popular belief, it is generally accepted that it was named after Marianne Fisher, the daughter of James Hurtle Fisher, the first resident commissioner. Spelling of names often varied in early colonial records depending on the educational standards of the early settlers. Marianne Fisher was born in 1827 and arrived in South Australia on board the *Buffalo* in 1836. She died in 1927 at the age of 100 years having outlived the rest of the *Buffalo's* company. Marion became a city in 1953. It was subdivided in 1838 by Light, Finiss & Co. for Henry Nixon & B. Finniss.

Marion Bay *SA* on Yorke Peninsula was named after the steamer *Marion* which was wrecked on a reef near the Althorpes in 1861 while under the command of Captain A. McCoy. The reef is now known as Marion Reef.

Marmion *WA* was named after Patrick Marmion who operated a whaling station in the vicinity in 1849. Marmion was declared a town-site under the Land Act on 5 April 1940.

Maroubra *NSW*, a locality in the Municipality of Randwick, is named, according to a journal dating from the First Fleet, after an Aboriginal called Marubrah who was a friend of Bennelong. Conjectured meanings of this name include 'like thunder' and 'good'.

Marree *SA* is a small town about 710 kilometres north of Adelaide. It is at the beginning of the famous Birdsville Track. Until the new line from Tarcoola to Alice Springs opened in 1980, Marree was a railway town serving the old Ghan line to Alice springs. 'Marree' comes from an Aboriginal word *marina* or *mari* meaning 'place of possums'. Dr N.B. Tindale says that the local tribe observed many possum tracks leading to the springs in ancient times. It was named on 20 December 1883 by the South Australian government. The area was previously known as Herrgott Springs after David Herrgott (sometimes misspelt Hergott and Herrgolt), a German collector and botanist, who was with Stuart Babbage and Tolimer and found the springs in 1859.

Marrickville *NSW* is a municipality in the inner metropolitan area of Sydney. Marrickville was the name given by Mr Thomas Chalder of St Peters to land he advertised for subdivision by auction on 12 February 1855. This area centred on the present day intersection of Victoria Road and Chapel Street, Marrickville. It was named after Mr Chalder's native Marrick, a very small village in Swaledale, Yorkshire.

Marryatville *SA* is a suburb of the City of Kensington and Norwood in the Adelaide metropolitan area. Marryatville was named by George Brunskill after Augusta Sophia Marryat. Marryat was the maiden name of the wife of Sir Henry Edward Fox Young, Governor of South Australia from 1848 to 1854. Lady Young's brother, Charles, was Dean of Adelaide from 1853 when he arrived in Adelaide. Another brother was Admiral J.H. Marryat, C.B. The first Bishop of Adelaide, Augustus Short, and Captain Frederick Marryat were uncles of Lady Young and Dean Marryat. Captain Marryat was the author of *Mr Midshipman Easy, Peter Simple, Masterman Ready,* and other stories. George Brunskill arrived in South Australia in1839 and leased this land from the South Australian Company. The old village developed in north-east corner.

Marsden Park *NSW* is a locality in the City of Blacktown on the western outskirts of Sydney. It is named after Samuel Marsden (1764-1836), Church of England chaplain, missionary and farmer.

Marsfield *NSW*, a northern Sydney suburban locality in Hornsby Shire and Ryde, gets its name from the Field of Mars which was once the name for the whole area. It was originally bestowed by Governor Phillip, but the reasons for naming it are obscure. He may have been thinking of Champ de Mars (the field of Mars) in Paris.

Martha, Mount *see* **Mount Martha**

Martin *WA* in Gosnells, a city in the Perth metropolitan area, is named after the Martin family, who helped pioneer the district. In particular, it is named after Edward Victor Martin (1901-1981), a long serving local council member. The name was suggested by the Gosnells Town Council in 1974.

Marulan *NSW* is an Aboriginal name which should be Murrawoolan, and may formerly have been spelt in this way.

Mary Kathleen *Q* was named by the Surveyor-General on 28

May 1956 after the wife of one of the discoverers of the uranium deposit.

Maryborough *Q* takes its name from the Mary River which the Aboriginals knew as the Moonabula. It was officially named the Mary River by government proclamation in 1847 in honour of the wife of the Governor of New South Wales, Sir Charles Fitzroy.

Maryborough *V* is a small city seventy kilometres north-west of Ballarat. It was originally named Simpson's Ranges after squatter brothers. Gold was discovered there in 1853 and Assistant Gold Commissioner J. Daly renamed it after his birthplace in Ireland.

Marysville *V* is a mountain resort township by Steavenson River thirty-five kilometres north-east of Healesville. It was named by workers in a water supply camp after Mary, wife of engineer John Steavenson.

Mascot *NSW* is a suburb of the Sydney municipality of Botany. It was known for many years as North Botany. By 1911, agitation by the people of North Botany for a distinctive name had reached the stage where they were granted a referendum. Three names were put forward—Mascot, Boronia and Booralee. Most people voted for Mascot and the municipal name was altered accordingly.

Matraville *NSW*, a locality in the Municipality of Randwick, comes from James Matra who sailed with Banks and Cook. He was an American who remained loyal to the British and his proposal to settle dispossessed British loyalists in New South Wales helped to interest the British Government in making a colony in Australia.

Mattingley, Mount *see* **Mount Mattingley**

Mawson *ACT* honours Sir Douglas Mawson (1882-1958), the Australian Antarctic explorer.

Mayfield *NSW*, a suburb of Newcastle, is named after May Scholey, one of the daughters of John Scholey, a Newcastle butcher who bought and subdivided the area in 1881.

Maylands *SA* was owned by Luke Michael Cullen and William Wadham when the township was laid out in 1876 and named after Mr Wadham's second wife, Emma Josephine May.

Maylands *WA* is a suburb in the City of Stirling in the Perth metropolitan area. Maylands was developed immediately following the opening of the Perth to Guildford Railway line. The Falkirk siding was constructed in 1899 to cater for Mr Mephan Pherguson and his foundry workers in manufacturing seamless pipes used in the Goldfields Water Supply Scheme. An increasing number of people settled in the vicinity of his factory, many being workers at the Midland Railway Workshops. In August 1899, a decision was made to build a station at the Falkirk Siding and on its completion in September, it was renamed Maylands. Mephan Pherguson's wife had died in 1893 and when he began his Western Australian operation his daughter Mary was his housekeeper. It is believed the name Maylands was coined soon after his arrival in 1898.

Meadowbank *NSW*, a locality in the Municipality of Ryde, was named after Captain Bennett's home Meadow Bank, near Charity Creek.

Meadows *SA* probably received its descriptive name from Charles Flaxman who applied for land in the area in 1839. The original survey, known as the 'Meadows Special Survey', was made in 1840. The first subdivision in the area that later became the town of Meadows was made by William Hall who purchased a section of the land in 1856.

Medina *WA*, the first of the residential suburbs of Kwinana, a town south of Fremantle, takes its name from the ship *Medina* which arrived in Western Australia on 6 July 1830 with 150 passengers. The name is derived from a river of the Isle of Wight.

Medindie *SA* is a suburb within the corporate town of Walkerville in the inner metropolitan area of Adelaide. According to Rodney Cockburn's *What's In A Name*, it is situated on land originally granted to Charlotte Hudson Beare and was formerly called Penny's section. In 1854, it was referred to as Medindi. Medindie was laid out and named by William Wadham as Medindee. No record remains of the meaning of the name.

Medlow Bath *NSW*, a locality in the Blue Mountains west of Sydney, was originally named Browns Siding when a railway platform was opened in 1880. This name was later changed to Medlow. Mark Foy owned a hotel in the town with a health spa close by. Bath was added to the name for this reason.

Meeandah *Q* is a locality in the Brisbane metropolitan area and gets its name from a corruption of the word 'meander'.

Meekatharra *WA* was gazetted as a town-site on 30 August 1901. The name was taken from the nearby Meekatharra Spring, which is recorded on an early plan dated July 1885. This name is of Aboriginal origin. It means 'a place of little water'.

Megalong Valley *NSW*, a locality in the Blue Mountains west of Sydney, was believed to be derived from the Aboriginal word *mega* meaning 'hand' because of the topography of the gullies and peninsulas being similar to the imprint of a hand.

Melba *ACT* is named after Dame Nellie Melba (1861-1931), the Australian opera singer.

Melbourne *V* was originally known as Dutigalla or Bearbrass. Sir Richard Bourke, Governor of New South Wales, selected it as the administrative centre of the district in 1837. He renamed it Melbourne after Lord Melbourne, who was then Prime Minister of England.

Melbourne Street Names Bourke Street is named after Sir Richard Bourke, Governor of New South Wales, who visited the Port Phillip settlement in 1837; Collins Street is named in honour of David Collins, who made an ill-fated attempt to found Victoria's first settlement at Sorrento in 1803; Flinders Street honours Matthew Flinders, who visited Port Phillip Bay in 1802; La Trobe Street is named after Charles Joseph La Trobe, the first governor of Victoria from 1851 to 1854; Lonsdale Street honours Captain William Lonsdale, Melbourne's first magistrate and Victoria's first colonial secretary; Russell Street is called after Lord John Russell, a British prime minister; Spencer Street was named by Governor Bourke after John Charles Spencer, an English politician; Swanston Street honours one of Melbourne's founders, Captain Charles Swanston, chairman of the Derwent Bank, Tasmania.

Melbourne, Port *see* **Port Melbourne**

Melrose *SA* Probably named after Melrose in Scotland; but there is a tradition that it was named by a surveyor who fell ill and was nursed back to health by the mother of George Melrose, a pastoral pioneer.

Melton *V* is a city about forty kilometres from Melbourne. It was first named Pennyroyal Creek. Melton is thought to have eventually been named by Mr G.W. Rusden, Clerk of the Legislative Council, who was a member of the famous hunt organized by Thomas Henry Pyke on his property in the 1840s. It is assumed that he named the village after the hunting ground of Melton Mowbray in Leicestershire.

Melville *WA* is a city in the Perth metropolitan area. The locality was named after nearby Melville Water, which was named by Captain James Stirling in March 1827, in honour of Robert Saunders Dundas, 2nd Viscount Melville, K.T., First Lord of the Admiralty.

Melville Island *NT* Named by P.P. King during his north coast survey in 1818 in honour of Viscount Melville, first Lord of the Admiralty. The island had been sighted in 1844 by Tasman who thought it was part of the mainland. A settlement was formed by J.G. Bremer, who took possession of the island on 26 February 1824. The Aboriginal name was Yermalner.

Menai *NSW* in the Sutherland Shire was originally called Bangor by Owen Jones, a Welshman and first settler in 1895. It was changed by the Postmaster-General in 1910 because of confusion with Bangor in Tasmania. It was named after the Menai Straits between the Welsh mainland and the Isle of Anglesey, opposite Bangor. Since the opening of the land west of the Woronora River and the atomic reactor at Lucas Heights, the name of Bangor has been applied to the eastern suburb of this subdivision, for it was here in 1895 that the original settlement was established. Menai is now the southern suburb.

Menindee *NSW* An Aboriginal name meaning 'egg yolk'. An earlier name was Laidley Ponds. A store was built here in 1859 and in 1862 it was chosen as the site of a town to be named Perry. Later it reverted to the Aboriginal name, first of Menindie, then Menindee. Another Aboriginal name for the locality was Williorara.

Mentone *V* is a suburb of Mordialloc in the Melbourne metropolitan area by Port Phillip Bay. A syndicate led by Sir Matthew Davies bought and subdivided land during the 1880s and named the district after the Mediterranean health resort. It was formerly known as Dover Slopes.

Merewether *NSW*, a locality in the Newcastle area, takes its name from Edward Christopher Merewether who arrived in the colony in 1838 as an aide-de-camp to Sir George Gipps who was Governor of New South Wales from 1838 to 1846. Merewether married the eldest daughter of the original landowner Dr James Mitchell, father of the founder of the Mitchell Library.

Merlynstone *V* is a suburb of the City of Coburg in the Melbourne metropolitan area. It was called after Merlyn, daughter of Captain Donald Stuart Vain, former councillor and land dweller.

Merredin *WA* is named from the merritt tree which was used by the Aboriginals to make spears. Literally translated Merrit-in (Merredin) means 'the place of Merritt'. The town-site was gazetted in 1891.

Merrijig *V* is a farming district on the Delatite River nineteen kilometres east of Mansfield. Its name is Aboriginal for 'well done'.

Merrilands *V* is a suburb of the City of Preston in the Melbourne metropolitan area. Merrilands comes from an Aboriginal word *merri* meaning 'stoney'. The estate was referred to as Merrilands, during sales of 1838 and 1839.

Merriwa *NSW* There have been several conjectures on the origin of this Aboriginal name. It seems that it was a favourite camping ground, and that the flat land at the junction of the Goulburn and Merriwa Creeks produced an abundance of grass. The seeds were gathered and ground to produce flour, while the name meant something like 'fertile place producing much grass seed.'

Merrylands *NSW*, a suburb of the Municipality of Holroyd in Sydney's western suburbs, was an estate where Arthur Todd Holroyd established a dairy. He named Merrylands Estate after a family property near Guildford, in England.

Mersey River *T* was discovered during a survey which had been organised by the Van Diemens Land Company and named by Edwin Curr after the English river. As it was the second river encountered after leaving Launceston, it was at first called Second Western River. The Aboriginal name was Paranaple.

Merton *V*, a pastoral district sixteen kilometres west of Bonnie Doon, is named after Merton Place in Surrey.

Middle Park *V*, a locality in South Melbourne, got its name because its railway station is mid-way along the south-western boundary of Albert Park.

Midland *WA* was named in 1894 by the Midland Railway Company which had been granted land on condition it linked up with the state railway line at Geraldton. In 1884, John Waddington suggested a railway be built between York and Geraldton, in return for grants of land on the route. In February 1886, the Midland Railway Company was forced to build the line between Guildford and Geraldton, choosing a site in the present town of Midland for its headquarters and workshops. The railway station at the junction of the two lines was named Midland Junction after the Company, but when a municipality was declared in 1895 it was named Helena Vale. The railway station kept its name and so much confusion arose that the town was renamed Midland Junction in 1901 and finally shortened to Midland in 1961.

Mildura *V* is a city and tourist resort on the Murray River, 544 kilometres north-west of Melbourne in the centre of a large wine district. Its name comes from an Aboriginal word meaning 'sore eyes' or 'red sand'.

Miles *Q* is a town in Murilla Shire on the Darling Downs. The explorer, Ludwig Leichhardt, named Dogwood Creek in 1844 on his way to Port Essington (now Darwin). The original settlement that began to form after the erection of a railway bridge in 1878 was called Dogwood Crossing. It was later named Miles in commemoration of William Miles (1817-1887) a Scots-born early landholder who was a member of parliament from 1865 to 1887.

Milingimbi *NT* is the site of the Methodist mission station. It takes its name from a large mythical snake which was supposed to inhabit the caverns at the Macassar Well.

Millicent *SA* is a town in south-eastern South Australia about 410 kilometres from Adelaide. The town was surveyed in 1870. It was named at the request of Governor Fergusson after Millicent Glen, the wife of George Glen, the owner of the Mayurra Station on which the town was sited. Mrs Glen, who was the daughter of Augustus Short, Bishop of Adelaide, was the only white woman (apart from her maids) on the station in 1854. Mayurra was derived from the Aboriginal word *maayera*, meaning 'fern straws'.

Millmerran *Q* derives from the Aboriginal words *meel* meaning 'eye', and *merran*, meaning 'to look out', because the site was used as a lookout by the Aboriginals.

Millswood *SA* is a suburb of the City of Unley which is located just south of central Adelaide. According to W.A. Norman's *History of Mitcham*, Millswood is named after Samuel Mills, a Scottish builder who arrived in South Australia in 1839 and settled in the Goodwood area. Mills called his property 'Ravenswood Farm' and the name has been perpetuated in the township of Ravenswood. According to Norman, Millswood Estate and Mills Street are named after this pioneer who donated the land where the Goodwood Institute now stands.

Rodney Cockburn's book *What's In A Name* says that Millswood embraces the name of George Mills, of Hill, Mills and Co., carriers who owned the section of land on which the suburb now stands and this explanation seems to be the most likely. Lands department records show that the land was originally granted to Thomas Hardy in 1838. It was subdivided in 1857 and called Goodwood Park. A plan of 1865 shows a portion of the land was owned by W. Raven and this may be the origin of the name Ravenswood. In 1868, the land was conveyed to George Mills from Silvester Percy

Badman. This area was subdivided in 1882 by D. Tweedie and G. Howell and called Millswood.

Milparinka *NSW* An Aboriginal term meaning 'water may be found here'. It was near here that a terrible drought caused Charles Sturt to remain for a period of six months.

Milperra *NSW* is a suburb of Bankstown, a city located in the western part of the Sydney metropolitan area. It is an Aboriginal word meaning 'company'.

Milsons Point *NSW*, a locality in North Sydney, is named after James Milson (1783-1872) who was one of the first white settlers in North Sydney.

Minmi *NSW* means 'home of the giant lily'.

Minnamurra *NSW* An Aboriginal word meaning 'plenty of fish'.

Mintabie *SA* There is no record of the origin of the name. The name of the Opal Fields settlement was taken from the nearby Mintabie Dam and yards located on the Granite Downs Pastoral Lease (Lease Number 2426). The name, Mintabie, was approved for the town by the Geographical Names Board on 27 October 1981.

Minto *NSW* was named by Macquarie after his friend Gilbert Kyngmount, Earl Minto, who served as Governor of India from 1807 to 1814.

Miranda *NSW* in the Sutherland Shire was named after Miranda, a character in the Shakespearian play *The Tempest*. According to James Murphy, manager of the Holt-Sutherland Company which he formed in 1881, he chose the name because he thought it 'euphonious, musical and an appropriate name for a beautiful place'.

Mirrabooka *WA* in Stirling, a city in the Perth metropolitan area, is an Aboriginal name for the Southern Cross.

Mitcham *SA* is a suburb and corporate city in the southern metropolitan area of Adelaide. Land for a village around Brown Hill Creek was surveyed and advertised in 1840. The name Mitcham was adopted by the manager of the South Australian Company William Giles, from his home town in Surrey, England. In 1853 the inhabitants of Unley and Mitcham formed the first district council of South Australia. Mitcham became a municipality in 1943 and a city in 1946.

Mitchell *ACT* recalls Sir Thomas Livingstone Mitchell (1792-1855), the Australian explorer, who opened up more of New South Wales and Victoria.

Mitchell *Q* is the main town of the Shire of Booringa about 600 kilometres west of Brisbane. It is named after the explorer, Sir Thomas Mitchell.

Mitchell River *NSW Q V* The rivers in these three states were all named after T.L. Mitchell, the well-known explorer who conferred many place names and was Surveyor-General of New South Wales. The Mitchell River in New South Wales was also known as the Mann River. In Queensland the Mitchell was named by Leichhardt on 16 June 1845, and in Victoria by Angus McMillan in 1840.

Mitchellstown *V* is a rural locality south-west of Nagambie. It was formerly known as The Old Crossing Place because it was the place where Major T.L. Mitchell crossed the Goulburn River in 1836. It was also known as Deegay Ponds. Mitchellton winery is located nearby.

Mitta Mitta *V* is a township at the junction of Mitta Mitta River and the Snowy Creek, 110 kilometres north of Omeo. Gold was discovered there in 1852. The name is Aboriginal meaning either 'little waters' or 'where reeds grow'.

Mittagong *NSW* An Aboriginal name meaning 'small mountain', 'high, rocky, scrubby hill' or 'plenty of dingoes'. The name did not take its present form officially until the railway came here in 1867. Governor Macquarie referred to it as Marragan and Minnikin in 1816, although by 1820 he spelt it Mittagong. The original Aboriginal name may have been Mirragong. In his autobiography, Martin Cash, who lived here and later became a bushranger, called it Meadow Gang. In 1849 a small setlement called Fitzroy was established near the ironworks, and in 1862 Crown land was subdivided and named the Village of Fitzroy. Another subdivision was named New Sheffield because of the association of the ironworks. The area was also known as Nattai.

Moama *NSW* was earlier known as Maiden's Punt, after the man who operated the ferry. It was to this point that Captain W.R. Randell brought the steamship *Mary Ann* and this proved to be the beginning of river trade on the Murray.

Moana Beach *SA* In November 1926 the Lake Beach Estate Limited asked the Nomenclature Committee to judge a competition that was being run to name the new seaside resort known at that time as Dodds Beach. The name, Boon Boona Beach, was originally selected, but because it was a slight variation in spelling of an Aboriginal word meaning 'the sea beach' the name, Moana, a Maori word meaning 'blue sea', was finally chosen as the winning name and Mr C.H. Cave the winning entrant.

Modbury *SA* is a suburb in the City of Tea Tree Gully in the north-eastern section of the Adelaide metropolitan area. Robert Symons Kelly encouraged the development of a village on his land which he named Modbury after his birthplace, Modbury in Devonshire. The village began to emerge between 1857-1858. Kelly purchased an 80 acre section (32 ha) at Upper Dry Creek in 1842.

He named his home 'Trehele' after another Trehele, a manor house south of Exeter in Devon. The name later became written as Treehill, but the property was known generally as Modbury Farm. It was the hotel to which the name Modbury was first applied and it is believed that it was almost certain that Kelly, a builder by trade, constructed the two-storey hotel.

In September 1858, William Stoneham applied to the district council for a general licence for a public house 'to be called the Modbury Hotel'. Stoneham became the first licensee but Robert Kelly retained ownership. The *South Australian Gazetteer* of 1866 describes Modbury as: 'a pastoral village . . . situated on the Dry Creek . . . and is in the midst of an agricultural country where wheat and hay are extensively grown. Modbury has a post office and one hotel — The Modbury. The resident magistrate is R.S. Kelly, Esq., J.P.' The Aboriginal name was Kirra Ung Dinga or Kirra Un Dunga.

Moe *V* is a city in Central Gippsland. The name is said to be an Aboriginal word meaning 'mud swamp'. It was formerly known as Westbury.

Molesworth *V* is a dairying district fourteen kilometres north-east of Yea. It was named by Surveyor Thomas Pinninger after Sir William Molesworth, chairman of the Transportation Committee which recommended the abolition of the transportation of convicts.

Moliagul *V* is a township on Burnt Creek, thirty-six kilometres north of Maryborough. Gold was discovered there in 1852. The name comes from an Aboriginal word meaning 'hill with trees'. It is pronounced 'Molly-EYE-gill'.

Molong *NSW* An Aboriginal term meaning 'all rocks'. Also the name of an estate in the vicinity.

Mona Vale *NSW* is a suburb of Warringah Shire in Sydney's northern beaches area. Mona Vale was named by a local farmer, J. Foley, after Mona Vale in Scotland.

Monaro; Monaro Plains; Monaro Range *ACT* An Aboriginal name, probably a corruption of Maneroo, meaning 'breasts' or 'navel'. Olaf Ruhen states the meaning is most probably breasts, for the hills, particularly towards the southern end, are breast-shaped, and have outcrops of granite for nipples at their peaks. There have been several spellings—Maneroo, Monaroo, Manaroo, and Menaroo. They were discovered in 1823 by Captain Mark Currie, accompanied by John Ovens and Joseph Wild. They named the plains Brisbane Downs after Sir Thomas Brisbane.

Monarto *SA* was planned during the 1970s as a new town in the area west of Murray Bridge in South Australia. Planners expected the 16 000 hectare site to have an eventual population of 200 000 but plans were deferred in 1976 because of lack of funds, and in 1980 the Government decided to abandon the Monarto project and sell the assets. It was said to be named after a Queen Monarto, wife of the River Murray Aboriginal tribe chief, King John. Its meaning is said to be 'thunder'. The Hundred of Monarto was proclaimed in 1847.

Monash *ACT* is named after General Sir John Monash (1865-1931), an outstanding Australian military commander in World War II.

Monbulk *V* is a locality in the Shire of Sherbrooke on the outskirts of Melbourne. Monbulk comes from the Monbulk Run established in 1838 and taken over by Thomas Dargon in 1850. Dargon heard the local Aboriginals use the words *mon bulk* and had them placed on the homestead gates. The words may refer to granite rocks which formed a useful hiding place on the hills. Monbulk Jam Factory, established by Daniel Camm in 1914, has made the name, Monbulk, familiar in all states of Australia.

Mont Albert *V* is a suburb of the City of Box Hill in the Melbourne metropolitan area. Mont Albert honours Prince Albert, the husband of Queen Victoria.

Montagu Island *NSW* After George Montagu Dunk, Earl of Halifax. The island was seen by Captain Cook, but it was not recognised as an island until seen from the convict ship *Surprize* in 1790. The Aboriginal name was Barunguba.

Montague *V* is a locality in South Melbourne. The name comes from Montague Street which in turn derived its name from a number of English families of that name living there.

Montebello Islands *WA* are a group of coral islands off the north-western coast of Australia. They were named by Nicolas Baudin in *Le Geographe* in 1803 in honour of

Marshall Lannes, who was elevated to the title of Duke of Montebello by Napoleon. The first British nuclear device was tested there in 1952.

Monterey *NSW*, a suburb of the Municipality of Rockdale, was named after the original subdivision.

Montmorency *V* is a locality in the Shire of Eltham on the outskirts of Melbourne. Montmorency was originally occupied by Captain Benjamin Baxter in 1840. The farm known as the Montmorency Estate eventually gave its name to the area.

Moolap *V* is a district eight kilometres south-east of Geelong. The Aboriginal word, *moo-laa*, means 'men gathering to go fishing'. The Aboriginal name for Point Henry, near Geelong, was Maloppio.

Moomba *SA* was chosen by Jeff Green, an American who was the resident manager of Australian Petroleum Ltd. The gas field was named after some dry claypans in the area which was actually known as Lake Moonba. Delhi altered it to Moomba because it was easier to pronounce. *Moomba*, acording to the people who chose it for the famous Melbourne festival, is an Aboriginal word meaning 'let's get together and have fun'. It is therefore a good choice for a festival. Unfortunately there is no record of any Aboriginal word with this meaning. Several other meanings have been found for *moomba* including 'big noise' and 'thunder'. The original Moomba remains in doubt. The nearest Aboriginal word anybody has found is *moonda* meaning 'beyond'.

Moonah *T* appears to have been applied in the mid 1890s. It is an Aboriginal word meaning 'gum tree'.

Moonee Ponds *V* is a suburb of the City of Essendon in the Melbourne metropolitan area. Moonee Ponds has been immortalised by Barry Humphries' character Dame Edna Everage. Moonee Ponds probably comes from an Aboriginal word but there is some dispute as to its meaning. Some authorities say it was used as a personal name, others claim it meant 'lizard'. A further explanation suggests it meant 'plenty of small flats'. The 1918 Victorian Railways list of stations says it was called after Moonee Moonee, an Aboriginal attached to the Mounted Police, who died in the Wimmera in 1845. Another explanation claims it was not Aboriginal at all but comes from John Moonee, a former British soldier who was granted crown land around the present site of Moonee Valley Racecourse.

Moonta *SA* is a corruption of an Aboriginal name, Moonta Moonterra, meaning 'place of impenetrable scrub'. The name was confirmed by the Governor, Sir Dominick Daly, when the town was laid out in 1863. The Moonta copper mine was discovered in 1861 by Patrick Ryan, a shepherd who worked for Sir Walter Watson Hughes. The Moonta mine was the first mine in Australia to pay a million pounds in dividends.

Moorabbin *V* is a city about ten kilometres south-east of the centre of Melbourne. The Bunurong tribe of Aboriginals, the first inhabitants of the area, have left the name Moorabbin, which means 'mother's milk'. It also means 'resting place' because the women of the tribe with young children used to rest in the district while the men went walkabout. The name, Moorabbin, was adopted by one of the cattle runs set up in the 1840s. Moorabbin became a district in 1862 and a shire in 1871. It was proclaimed a city in 1934. Moorabbin airport opened in 1949.

Moore Park *NSW* is a locality in the City of Sydney. It is named after the Mayor of Sydney who established the council's right to the land in the 1860s. It contains two cricket grounds, two athletic tracks, the Royal Agricultural Society's showground and a golf course. Beneath the showground runs Sydney's second water supply, Busby's Bore, a tunnel hacked out by convict labour between 1827 and 1837 under the supervision of John Busby (1765-1857), a colonial surveyor and engineer.

Moorebank *NSW*, a suburb of Liverpool takes its name from the principal landholder of the area in the early days of settlement. Thomas Moore first came to New South Wales in 1792 on board the *Britannia* as ship's carpenter. In 1796 he was appointed government boat builder, a position he held until 1809 when he resigned to settle on land he owned on the right bank of the Georges River. He built himself a substantial brick home where he entertained Governor Macquarie during his visit to the district in 1810. Unfortunately, the home was demolished about 1940.

Moorefields *NSW* is a suburb of the Sydney municipality of Canterbury. Research may find the correct origin of this name from a number of possibilities: 1. Patrick Moore owned land near King Georges Road and had a 60 acre (24 hectare) grant at Kogarah which was called Moorefield; 2. W.H. Moore, Acting Attorney-General in 1826, had extensive land-holdings in the vicinity; 3. Moorfield, near London, was a popular preaching place of John Wesley and the many Wesleyans in the area adopted the name. While present day spelling is Moorefields, early reference was usually to Moorfield; 4. The Chard family may have suggested the name Moorfield or called part of their farm 'Moorfield', just as another part was named 'Taunton Farm'. Their place of origin was Taunton, Somerset, and there was a Moor Farm nearby.

Moorooduc *V* was the original name for Schnapper Point. Its meaning is uncertain but it may come from the Aboriginal word *murraduk* meaning 'flat swamp'.

Mooroolbark *V* comes from an Aboriginal word meaning 'red earth'.

Morang *V* comes from an Aboriginal word meaning 'cloudy'.

Mordialloc *V* is a city and popular seaside resort in the Melbourne metropolitan area on the shores of Port Phillip Bay about twenty-five kilometres from the centre of Melbourne. Mordialloc probably came from the originally Aboriginal name for the shallow, muddy creek which deepened before it entered the bay. The first white settlers took up land in the 1840s. O'Shannassy's Windert cattle run covered an area from Point Ormond to Mordialloc. Local government in the area began in 1862. It became a city in 1926.

Moree *NSW* is known as 'the inland health resort' because of its artesian waters. These waters are believed to have therapeutic value in the relief of rheumatic ailments. They have a temperature of about 45°C and a high mineral content. The name, Moree, comes from an Aboriginal word meaning 'spring' or 'waterhole'. The town was surveyed in 1859 and proclaimed in 1862.

Moreland *V* is a suburb of the City of Coburg in the Melbourne metropolitan area. Moreland was named by Dr Farquahar McCrae, an army surgeon who was one of the district's pioneers. Moreland was the sugar plantation in Jamaica on which his father had been born.

Moreton Bay *Q* was named Morton Bay on 17 May 1770, in honour of James Douglas, Earl of Morton, who was president of the Royal Society. The present spelling comes from a wrong spelling in John Hawkesworth's account of Cook's voyages, published in 1773.

Morgan *SA* The town was proclaimed 1878, and named after Sir William Morgan, Chief Secretary of South Australia, and later Premier.

Morgan, Mount *see* **Mount Morgan**

Morley *WA* began appearing on maps around the turn of the twentieth century and was adopted when the area was subdivided for urban development after World War II. It is thought to be named after Charles William Morley who was born on his father's Upper Swan farm on 20 July 1840. He is known to have farmed in the Morley area during the 1860s and 1870s.

Mornington Peninsula *V*, is a stretch of land running from Frankston to Portsea on the east side of Port Phillip Bay. Mornington was named after the Earl of Mornington.

Morpeth *NSW* was named after an English town which is fourteen kilometres north of Newcastle. The name is Old English for 'murder path'. The area was originally named Green Hills in 1801, by Lieutenant-Colonel Paterson. It was changed to Morpeth in 1834.

Morphett Vale *SA* was named after John Morphett, a member of the first exploration party in 1837. He arrived in South Australia in the *Cygnet* on 11 September 1836 and was present at the proclamation of the province under the Old Gum Tree at Glenelg. He became a member of the Legislative Council in 1843, and later became Sir John Morphett.

Morphettville *SA* is a suburb within the corporate city of Marion, a suburban city in the Adelaide metropolitan area. Morphettville commemorates Sir John Morphett (1809-1892), an early supporter of the scheme to settle South Australia in September 1836. Two months later, with Lieutenant Field and G.S. Kingston, he discovered the River Torrens. He considered Kangaroo Island unsuitable, and his votes to move the settlement to the mainland were decisive. In Adelaide, he became a land agent acting for English friends. He threw himself behind every good cause. In 1840 he became treasurer of Adelaide's Municipal Corporation. He helped to found the agricultural society in 1844, and gave support to the Collegiate School of St Peter. He also had a long and distinguished political career. In 1843, he became one of the first non-official nominees in the Legislative Council. When the Legislative Council was reformed in 1851, he was elected speaker. He was president of the Legislative Council from 1865 to 1873.

Morrill *Q* honours James Morrill, a seaman from Maldon in Essex, who was one of seven people who, after incredible hardships, drifted ashore on a raft at Cape Cleveland, forty-two days after his ship, the *Peruvian*, was wrecked on Horeshoe Reef. Those who escaped from the sea died, one after the other, in the bush. Morrill became the last survivor of that disaster. For seventeen years, from 1846 to 1863, he lived with the Aboriginals, always hoping that one day he would be able to return to the life and the people he knew. Eventually, a sick man, he made contact with two men working on Antill's Station near the Burdekin River, and was taken to Bowen. Morrill was granted one of the first quarter-acre blocks of land in Townsville, at the lowest concession price of only four pounds. Unfortunately he died in his early forties, less than three years after his return to civilisation.

Morrisons Bay *NSW*, a locality in the Municipality of Ryde, was on the landholding granted in 1795 to Archibald Morrison, a private in the New South Wales Corps.

Morrisset *NSW* was named after Major James Morrisset (1780-1852) who was the military commander and magistrate at Newcastle in 1818, at Bathurst in 1823 and Norfolk Island in 1826.

Mortdale *NSW*, a suburb of the Municipalities of Kogarah and Hurstville, is named after Thomas Sutcliffe Mort, the pioneer merchant who was the father of the refrigerated meat trade in Australia. This was farming land which Mort bought as an investment.

Mortlake *NSW* is a locality in the Sydney municipality of Concord. It may have been named after Mortlake in Surrey. It was the scene of rowing races, an extremely popular sport in the 1880s, and Mortlake, like the names of Henley and Putney, may have been transferred from the Thames to Parramatta River about this time.

Mortlake *V* is a township by Mount Shadwell, fifty kilometres north-east of Warrnambool. It was named after Mortlake in Surrey meaning 'young salmon stream'.

Morwell *V*, a large town in central Gippsland, was named by C.J. Tyers, Lands Commissioner of Gippsland in 1884, and is thought to have been derived from the town of Morwellham and a former Abbey in Devon, England.

Mosman *NSW* is a municipality on Sydney's North Shore. Mosman derives its name from Archibald Mossman (1799-1863) who set up a whaling station at Mossman's Bay in 1830. He built a stone wharf, a large residence for himself, and two stone dwellings for ships' officers and crews. Mossman sold out in 1839 and moved to the New England district. He is buried at Randwick. The whaling industry, which provided half the income of New South Wales in 1838, collapsed in 1851.

Mosman Park *WA* is a town on the Swan River in the Perth metropolitan area. It was called Buckland Hill until 1937 when it was changed to Mosman Park after Mosman in Sydney, the birthplace of Road Board member, R.J. Yeldon.

Mount Arapiles and **Arapiles** *V* is a farming locality approximately thirty kilometres west of Horsham. Mount Arapiles was named by Major T.L. Mitchell in 1836 after the village in Spain where the Battle of Salamanca was fought.

Mount Augustus *WA*, the largest rock in the world, was climbed by the surveyor, F.T. Gregory, in 1858 and named after his brother Augustus who was then leading an expedition in search of the remains of the German explorer Ludwig Leichhardt.

Mount Barker *SA* is a town about thirty-four kilometres south-east of Adelaide. It is located about five kilometres from the mountain of the same name in the Mount Lofty Ranges. The mountain was named by Charles Sturt in honour of his friend, Captain Collet Barker, of the 39th Regiment, who was killed by Aboriginals on 30 April 1831 near Goolwa.

Mount Barker *WA* is called after the nearby mountain which was named by Surgeon T.B. Wilson RN in 1829 in honour of Captain Collet Barker who was then in command of the King George Sound outpost. The town was named by Doctor Alexander Collie in May 1831.

Mount Bartle Frere *Q*, the highest peak in Queensland, rises to 1611 metres above sea level. It was named by the explorer, George Elphinstone Dalrymple, in 1873 in honour of Sir Henry Bartle Frere, who was then president of the Royal Geographical Society in London.

Mount Bellenden Ker *Q* is located in Bellenden Ker Range. Captain Philip Parker King named the range in 1819 after John Bellenden Ker, a botanist of the period.

Mount Bischoff *T* was named after James Bischoff, the second managing director of the Van Diemens Land Company.

Mount Blackwood *Q* recalls the visit of Captain Blackwood of the Royal Navy. In March 1842 he landed near the present day location of Mackay Outer Harbour in search of water.

Mount Bogong *V*, an alpine peak east of Mount Beauty township, gets its name from the Bogong moth, a delicacy favoured by the Aboriginals.

Mount Bruce *WA* was named after Lieutenant Colonel John Bruce, who was then commandant of troops supporting the expedition by the explorer, Francis T. Gregory, who discovered it in July 1861. The 1227 metre peak is within the rich iron ore deposits area of the Pilbara region.

Mount Buffalo *V*, which is located about seventy kilometres south-east of Wangaratta, was named by the explorers, Hume and Hovell, during their expedition from Sydney to Port Phillip Bay in 1824. The mountain looks like a buffalo from one vantage point.

Mount Buller *V* which is located forty-eight kilometres east of Mansfield was named by T.L. Mitchell in 1835. He called it after a friend in the Colonial Office. The Aboriginal name was Marrang.

Mount Canobolas *NSW* rises 1395 metres about eighteen kilometres from the city of Orange. Its name is derived from two Aboriginal words, *coona booloo* meaning 'two heads' or 'twin shoulders'. The mountain has two main peaks. The explorer, George Evans, sighted Mount Canobolas in 1813 and the explorer Sir Thomas Mitchell climbed it in 1833.

Mount Colah *NSW*, a locality in Hornsby Shire on the North Shore of Sydney, is derived from the same source as the Aboriginal word for 'koala'.

Mount Coot-tha *Q* is a locality in the Brisbane metropolitan area and overlooks Brisbane. The name was suggested by A.W. Radford in 1883. It is an Aboriginal word meaning 'place of native honey'. When the hill was cleared of heavy bush in the early days of settlement, to provide a look-out above Brisbane, one tree was left standing, and for this reason the hill was known for many years as One Tree Hill.

Mount Cotteril *V*, a hill in a rural locality south of Melton, is named after A. Cotteril a member of John Batman's Port Phillip Association.

Mount Cotton *Q*, a town in the Brisbane area, was named after Major Sydney Cotton by Surveyor Dickson in 1840.

Mount Daintree *Q*, was named by William Hann who led the Northern Expedition Party of 1872, after Richard Daintree. Hann and his brother Frank were partners with Daintree at Maryvale Station on the Clarke River from 1864. Daintree worked as a geologist and petrologist in Victoria. In 1869 he was appointed Government Geologist for Northern Queensland and then became Agent-General in London.

Mount Dangar *NSW* is named after Henry Dangar, surveyor to the Australian Agricultural Company.

Mount Diogenes *V*, a hill north of Mount Macedon, is more commonly known as Hanging Rock. It was named after the fifth century BC Greek philosopher.

Mount Disappointment *V*, a peak of about fifty kilometres north of Melbourne near Whittlesea, was named by Hume and Hovell, who had hoped to see the ocean when they ascended the Plenty Range in 1824.

Mount Druitt *NSW*, a locality in the City of Blacktown on the western outskirts of Sydney, was named by Major George Druitt (1775-1842), military officer, public servant and settler, who was granted land in the area by Governor Macquarie.

Mount Eliza *V*, a hill and residential area south of Frankston, was named by William Hobson of HMS *Rattlesnake* in 1836 after John Batman's wife, who was later killed in a drunken brawl at Geelong.

Mount Field National Park *T* is named after Baron Field, a judge of the New South Wales Supreme Court, who visited Tasmania in1819 and 1821 as an itinerant judge. It is Tasmania's first national park. A reserve was first proclaimed in 1885.

Mount Gambier *SA* was sighted on 3 December 1800 by Lieutenant James Grant in HMS *Lady Nelson.* He named Mount Gambier after Admiral Lord James Gambier. It was his first sight of Australia en route to Sydney from England. According to Rodney Cockburn's book *What's in a Name*, the site of the town was originally owned by Evelyn Sturt, brother of Charles Sturt, who issued leases to traders who had erected premises some time prior to 1849. The leases were over allotments that were later described in a plan laid out as Gambier Town (private town), and deposited in 1856 by Hastings Cunningham, to whom Sturt had conveyed his interest in the land in 1854. The area became known as Mount Gambier, and the subdivision in 1880 of two allotments within the private town of Mount Gambier were laid out as Gambierton (Deposited Plan 1306). The government town of Gambier Town was surveyed in1861 by A.H. Smith. The name was altered to Gambiertown in the *Government Gazette* dated 5 April 1879. The government town of Mount Gambier was surveyed and proclaimed in1951. The Aborigines knew the locality as Ereng Balam or Egree Belum, 'place of the eagle-hawk'.

Mount Gravatt *Q*, is a locality in the Brisbane metropolitan area and is named after Lieutenant George Gravatt, who was in charge of the Moreton Bay settlement for three months in 1839.

Mount Hawthorn *WA* was unoccupied for the first fifty years after the founding of the Colony of Western Australia. In the late 1890s it was bought by a syndicate which included James Hicks. When this group subdivided their land in 1903, James Hicks called his portion of the subdivision Hawthorn Estate, as he had recently been in Melbourne and stayed in Hawthorn. Residents later were apparently dissatisfied with the name. As parts of it were on an eminence and names with a 'Mount' prefix were fashionable at that time (e.g. Mount Lawley), they called it Mount Hawthorn.

Mount Hopeless *SA,* is the place where E.J. Eyre was finally convinced that his search for good, fertile country was hopeless. This was in 1840 on his third expedition. Even before climbing it, he was resolved 'to waste no more time or energy on so desolate and forbidding a reason'.

Mount Hotham *V,* a peak fifty-six kilometres south-east of Bright, was visited by Baron von Mueller, the noted botanist, during a botanical expedition in 1854. He named it after Sir Charles Hotham, who was then Governor of Victoria.

Mount Irvine *NSW,* a locality in the Blue Mountains west of Sydney, was surveyed in the late nineteenth century by Mr C.R. Scrivener, Surveyor-General, who applied the name because the area was referred to by the Lands Department as Parish of Irvine.

Mount Isa *Q* was named by John Campbell Miles, a bushman and prospector who made an accidental discovery of a rich silver-lead ore in 1923. He named the site Mount Isa after his sister Isabella.

Mount Jukes *Q* is named after a botanist in Captain Blackwood's party of exploration which landed near the present location of Mackay Outer Harbour in search of water in 1842.

Mount Kaputar *NSW* This name is of Aboriginal origin, being a corruption of Kapular, entered by T.L. Mitchell on a sketch map on 3 January 1832.

Mount Keira *NSW* is an Aboriginal name for 'brush turkey'.

Mount Kosciusko *NSW* is called after Tadeusz Kosciusko, the Polish patriot, and named on 15 February by Paul Edmund Strzelecki during his exploration of the Australian Alps in 1840. He wrote: 'The particular configuration of this eminence struck me so forcibly by the similarity it bears to a tumulus elevated in Krakow over the tomb of the patriot Kosciusko, that, although in a foreign country, but amongst a free people...I could not refrain from giving it the name Mount Kosciusko.' Strzelecki's memory was at fault. Kosciusko was in fact buried in Krakow cathedral, but a memorial mound was erected on the outskirts of the city. There has been much discussion on whether the peak Strzelecki climbed was actually Kosciusko. It may have been Australia's second highest peak which is nearby Mount Townsend. The Aboriginal name of this part of the Australian Alps was Muniong or Munyong. Most Australians pronounce the mountain as Kos-ee-OS-ko although a few locals still call it Koz-ee-US-ko. The Polish pronunciation is more like KerSHISH-ko.

Mount Ku-ring-gai *NSW,* a locality in Hornsby Shire on the North Shore of Sydney, is named after the Ku-ring-gai Aboriginal tribe. It may mean 'the hunting grounds of the Kuring clan'.

Mount Lawley *WA* in Stirling, a city in the Perth Metropolitan area, was named after Sir Arthur Lawley who was Governor of Western Australia in 1901-1902. A distinguished soldier, he spent only fifteen months as Governor. Most of his colonial service was spent in South Africa.

Mount Lofty *SA* While Matthew Flinders was examining the St Vincent Gulf in the *Investigator,* he used the words 'a lofty mountain' in his *Rough Log* on 23 March 1802, and on 1 April, 'Mount Lofty'. The latter date is therefore to be regarded as that on which the name was conferred. A monument on the summit commemorates the sighting and naming. The Aboriginal name was Yure Idla 'place of the ears', or 'whelp's ears'.

Mount Lyell *T* was discovered in 1862 by a geologist named Charles Gould, who called it after the famous British geologist, Sir Charles Lyell.

Mount Macedon *V* is the name of a mountain and nearby township seventy kilometres north-west of Melbourne. The mountain was sighted by Hume and Hovell in 1824 and climbed by T.L. Mitchell on 30 September 1836. Mitchell could see Port Phillip from the summit, and by recalling his ancient history named it Mount Macedon, apparently because Philip of Macedon, the father of Alexander the Great, happened to have the same name. The Aboriginal name was Geboor.

Mount Martha *V*, a hill and holiday resort on Mornington Peninsula, was named in 1837 after the wife of Captain Lonsdale.

Mount Mattingley *V* is named in honour of A.H.E. Mattingley, a naturalist who pressured the Government to reserve Wyperfield as a national park.

Mount Morgan *Q* was discovered by two pastoralists, Charles and William Archer in 1853. They named it Ironstone Mountain. It became known as Mount Morgan after 1882, when the brothers Thomas, Edwin and Frederick Morgan discovered gold there.

Mount Nelson *T* was originally applied to the mountain feature by Governor Lachlan Macquarie in 1811, in recognition of His Majesty's brig, *Lady Nelson*, which served in the pioneer landings at Risdon and Sullivans Coves.

Mount Newman *WA* is a peak 1053 metres above sea level in one of the main iron ore producing regions of the Pilbara. The landmark was named in 1896 by a mapping expedition party and recorded by the then leader, Surveyor W.F. Rudall. He called it after the thirty-year-old leader Aubrey Woodward Newman who had died of typhoid fever a few months earlier.

Mount Olga *NT* was discovered by Ernest Giles in 1872 and named by Baron von Mueller after Queen Olga of Württemberg, 'one of the royal patrons of science'.

Mount Ossa *T*, 1617 metres high, is the highest mountain in Tasmania. It is situated about midway between Cradle Mountain and Lake St Clair within that National Park. As the name first appears on a map of government geologist, Charles Gould, in the 1860s, it is assumed that he gave it, no doubt after the mountain in Greece which figures in Greek mythology.

Mount Pleasant *WA* was named by James Herbert Simpson who took up land in the area in 1911. Simpson built a house there which, because of its shape, became known as 'the castle on the hill'. Simpson named it Mount Pleasant, the name which now adorns the suburb.

Mount Pritchard *NSW* is a suburb of Fairfield on the south-western outskirts of the Sydney metropolitan area. Mount Pritchard was called after an estate agent named Pritchard who subdivided a large tract of land to the west of Cabramatta.

Mount Riverview *NSW*, a locality in the Blue Mountains west of Sydney, was formerly part of Blaxland and named by the Blue Mountains City Council in the *Government Gazette* on 24 April 1964, following a request by the local progress association. For many decades beforehand, a popular vantage point was known as Mount Riverview Lookout.

Mount Roland *T* is named after Captain Rolland of the 3rd Buffs Regiment, who explored the area in 1823.

Mount Rumney *T* perpetuates the name of William Rumney, the builder of 'Acton', one of the historic homesteads of south-east Tasmania. Rumney received a grant of land in the Clarence Plains district and this included the mountain that now bears his name.

Mount Tomah *NSW* is a locality in the Blue Mountains west of Sydney. It was originally called Mount Harrington by Lieutenant-Colonel Paterson of the New South Wales Corps when he made an unsuccessful attempt to cross the Blue Mountains. The name was later replaced by Tomah.

Mount Victoria *NSW* is a locality in the Blue Mountains west of Sydney. It was known in the early nineteenth century as One Tree Hill and Broughton's Water Hole. When the railway was constructed in May 1868, Governor Belmore felt that the name Victoria, in honour of the then Queen, would be appreciated and to avoid confusion with the adjoining state, the name Mount Victoria was applied.

Mount Warning *NSW* was so named in 1770 by Captain Cook because after passing it the *Endeavour* was carried perilously close to the rocky shoals of Point Danger.

Mount Waverley *V* is a suburb of the City of Waverley in the Melbourne metropolitan area. Mount Waverley was originally known as Waverley. It was probably named by Dr James Silverman who made the first subdivision of land in the district in 1854 and called it The Waverley Estate.

Mount Wellington *T* was known by a number of names until 1824 when it was given its present name in honour of the Duke of Wellington by Governor Sorell.

Mount Wilson *NSW,* a locality in the Blue Mountains west of Sydney, was surveyed in 1868 and named in honour of the then Minister for Lands, Mr J. Bowie Wilson.

Mount Wingen *see* **Burning Mountain**

Mount Woodroffe *SA* is located in the Musgrave Ranges near the north-eastern border of South Australia. It is 1440 metres above sea level and is the highest point in the state. It was first sighted by William Christie Gosse in 1873 from the summit of Ayers Rock, 130 kilometres to the north-east. It was named in honour of the Surveyor-General of South Australia, John Woodroffe Goyder.

Mount York *NSW,* a locality in the Blue Mountains west of Sydney, was named by Blaxland, Lawson and Wentworth on their attempt to cross the Blue Mountains in 1813. It was from there that they saw the fertile plains spreading westward. The name was confirmed by Governor Macquarie in 1815.

Mount Zeil *NT* was discovered by Ernest Giles in 1877 and named by von Mueller while editing Giles's *Geographic Travels in Central Australia.* It is believed that the origin of the name lies with Count Waldburg Zeil, who accompanied Baron von Heuglin to Spitzbergen in 1870 and made natural history collections there. Giles discovered the mountains including this one (the highest in the Northern Territory) and von Mueller commemorated his German colleagues through this place name.

Mudgee *NSW* is an Aboriginal word usually translated as 'nest in the hills'.

Mulgoa *NSW,* a locality within the boundaries, of the City of Penrith, was the Aboriginal name for Penrith.

Mullumbimby *NSW* is an Aboriginal word, possibly meaning small round hills. The original name may have been Mullibumbi.

Mundaring *WA* is a shire in the Perth metropolitan area. Mundaring is derived from an Aboriginal word meaning 'a high place on a high place'. A London immigrant, Peter Gugeri, established a vineyard in 1893 and renamed it 'Mundaring'. This name was suggested for the town-site by the surveyor responsible for its design in 1898. The local authority was originally called the Greenmount Road Board. The name was changed to Mundaring Road Board in 1934. This became the Shire of Mundaring in 1961.

Mundubbera *Q* derives from the Aboriginal words munda, 'a foot' and burra, 'a step', because the Aboriginals made cuts in the trees there to make climbing easier.

Mungo, Lake *see* **Lake Mungo**

Munmorah *NSW* means a 'quiet place'.

Munno Para is a district north of the Adelaide metropolitan area. The area is now largely subdivided and Munno Para was gazetted as a suburb of Adelaide in 1978. The name, meaning 'golden wattle creek', originated in the language of the Kaurna people, a now-extinct Aboriginal group.

Murchison *V,* a township on Goulburn River thirty-four kilometres south-west of Shepparton, was named after an early settler, Captain John Murchison.

Murchison River *WA* was named after Sir Roderick Murchison, a British scientist and secretary of the Royal Geographical Society, by George Grey during his heroic forced march from Shark Bay to the Swan River in 1839.

Murgon *Q* is an Aboriginal word for a type of water lily which grows in the district.

Murrarie *Q* is a locality in the Brisbane metropolitan area and was named by the Railways Department in 1907. It is an Aboriginal word meaning 'plenty' or 'sweet water'.

Murray Bridge *SA,* as its name implies, is located near a bridge over the Murray River, which was originally opened in 1879. The town was proclaimed on 6 March 1884 and named Mobilong. It was altered on 19 September 1940 to Murray Bridge to conform with the name of the railway station.

Murray River *NSW SA V* was first visited by white men when the explorers, Hamilton Hume and William Hovell crossed the river at Albury where the Hume Dam now stands on 16 November 1824. Hovell named it Hume's River because Hume was the first to see it. However, in 1830, Charles Sturt, unaware that it was the same stream, named it after Sir George Murray who was then Colonial Secretary.

Murrumbeena *V* is a suburb of the City of Caulfield in the Melbourne metropolitan area. *Murrumbeena is* an Aboriginal word that may mean 'belonging to you', 'welcome' or even 'place of frogs'.

Murrumbidgee River *NSW* An Aboriginal name meaning 'big water'. An unusual sequence occurred in connection with the name for it was known to white people before the river was discovered. In 1820 Charles Throsby, the explorer, informed the Governor that he was looking forward to finding 'a considerable river of salt water (except at very wet seasons), called by the natives Mur-rumbid-gee'. He reached the river in April 1821. Brigade-Major Ovens and Captain Mark Currie also saw it in 1823. The river was more thoroughly explored by Charles Sturt in 1829.

Murrurundi *NSW* A contraction of the Aboriginal name, Murrumdoorandi, 'place where the mists sit'. Other meanings given are 'mountain', and 'five fingers'.

Murwillumbah *NSW* is an Aboriginal name for which two possible meanings have been given. One is 'place of many possums', the other is 'camping place'. The town was surveyed in 1872.

Muswellbrook *NSW* was called Muscle Brook when it was first gazetted in 1833 and the name was not officially changed until 1949.

Myall Lakes *NSW* got their name from the Aboriginal word *myall* which was a local name for a small, silver-grey wattle.

Myrtleford *V,* is a town forty-six kilometres south-east of Wangaratta near the junction of the Buffalo and Ovens

Rivers. Henry Davidson surveyed the township site in 1856 and recommended the name Myrtleford because of the large number of myrtle trees growing in the district. Gold was discovered there in 1853.

Nagambie *V* is a township by Lake Nagambie twenty-seven kilometres north of Seymour. The name comes from an Aboriginal word *nogambie*, meaning 'lagoon'.

Nambour *Q* was named after 'Nambour' cattle station. *Nambour* is an Aboriginal word meaning 'red-flowering tea tree'.

Namoi River *NSW* has two possible derivations. One is a corruption of the Aboriginal word *nynamu*, 'breast', so called because the curve of the river resembled that of a woman's breast; or *ngnamai* or *njamai*, place of the myamai tree, a variety of wattle.

Nanango *Q* comes from an Aboriginal word whose meaning is unknown. It was formerly Goode's Inn.

Nangwarry *SA* is a sawmilling town 400 kilometres south-east of Adelaide. The name is derived from the Aboriginal word *ngrangware*, meaning 'path to the cave'.

Nar Nar Goon *V* is a township in West Gippsland, thirty-three kilometres northwest of Warragul. Nar-Nar-Goon is an Aboriginal word meaning 'native bear' (koala).

Naracoorte *SA* is believed to be derived from *narcoot* or *gnangekuri*, an Aboriginal word for 'big waterhole'. The government town was surveyed in January 1859. Naracoorte was originally spelt 'Narracoorte', but government departments dropped the extra 'r'.

Narellan *NSW* One account says that it comes from the name of a property belonging to Francis Mowatt; another that it was derived from 'Naralling', a property owned by William Hovell. In either case it is doubtless of Aboriginal origin.

Naremburn *NSW*, a locality in the Municipality of Willoughby, was known in the early days as General Township and is said by many local residents to have derived its name from an Aboriginal name.

Narooma *NSW* Of Aboriginal origin, this name is a corruption of 'Noorooma', the name of a cattle station.

Narrabeen *NSW* is a lake and a suburb in Warringah Shire in Sydney's northern beach area. The origin of its name is something of a puzzle. One theory is that it is from narrow beans which grew in the area and caused Dr White to become ill while travelling with Governor Phillip's party towards the coast on 22 August 1788. The name Narrobine Creek appears in records relating to two escaped convicts in the district in 1801 so it was in use at that time. Surveyor James Meehan placed the name Narabang Narabang Lagoons on his map in 1814. He claimed the words were of Aboriginal origin and referred to the number of swans in the area.

Narrabri *NSW* comes from an Aboriginal word meaning 'forks', 'forked sticks', or possibly 'large creek'. Other forms of the name are Nurraburai, Nurraburi and Nurruby, the latter being the name of the station taken up by Patrick Quinn and Cyrus M. Doyle in 1834.

Narrabundah *ACT* has been an Aboriginal place name associated with the locality since the days of the early settlers.

Narrandera *NSW* An Aboriginal name meaning 'place of lizards'. It was the name of a pastoral holding and was gazetted a village in 1863.

Narraweena *NSW* is a suburb of Warringah Shire in Sydney's northern beaches area. Narraweena is an Aboriginal word meaning 'a quiet place on the hills'. It was given to the area after World War II when a subdivision was made to build Housing Commission homes.

Narre Warren *V* is a suburb of the City of Berwick on the outskirts of the Melbourne metropolitan area. Narre Warren was named in 1841 when a pioneer named Daniel Taylor lived there. The name could be a corruption of Aboriginal words *daana warree* meaning 'holly country' but another translation *narre woran* meaning 'no good water' is also possible. *Narre* meant 'she-oak' and *waran* meant 'water' so that the name may have meant that 'creek water by she-oaks is no good'.

Narrogin *WA* is a corruption of the Aboriginal words *gnarajin* or *gnargajin,* which mean 'place of water', or 'waterhole'.

Narromine *NSW* is derived from an Aboriginal word *gnarrowmine,* meaning 'place for honey', or *gnaroomine,* meaning 'bony Aboriginal man'. 'Narrowmine' cattle station was established by Thomas Raine in 1840.

Narwee *NSW,* a suburb of the municipalities of Hurstville and Canterbury, was the name given by the Department of

Railways to the new station when the East Hills railway line was opened on 21 December 1931. It is an Aboriginal word for the sun (the railway station at Jannali was also named in 1931, the name being an Aboriginal word for the moon).

Naturaliste, Cape *see* **Cape Naturaliste**

Nedlands *WA* was part of land bought in 1854 by Colonel John Bruce, then Military Commandant of the Colony. He intended the land to be the heritage of his son, Edward. The estate was therefore referred to as 'Ned's Land'. Nedlands was proclaimed a city on 5 June 1959.

Neerim *V* is a farming district twenty-seven kilometres north of Warragul. Its name comes from an Aboriginal word meaning 'war spear'.

Nelson, Mount *see* **Mount Nelson**

Nelson Bay *NSW* is the name of two inlets on the east coast of Australia.
    1. Nelson Bay is located in the Sydney suburb of Waverley. Nelson Bay was used as a name on the original plans of the grants to Mortimer Lewis, six years before Robert Lowe came to Sydney. The connection between Nelson Bay and Bronte is that after Admiral Nelson's victory at Aboukir in 1798, the King of Sicily created Nelson Duke of Bronte and gave him estates at Bronte in Sicily.
    2. Nelson Bay at Port Stephens was referred to by Sir Edward Parry, Australian Agricultural Company commissioner, as Nelson's Bay. The origin of the name is controversial. It was possibly named in tribute to Admiral Sir Horatio Nelson or perhaps in memory of the boat the *Lady Nelson,* in which Governor Macquarie travelled to Port Stephens.

Nepean River *NSW* flows within the boundaries of the City of Penrith. It was named by Governor Phillip after Sir Evan Nepean, Under-Secretary for the Home Department and Secretary to the Admiralty, after it was discovered by Watkin Tench and his companions in 1789.

Neutral Bay *NSW,* a locality in North Sydney, got its name in 1789 when England was at war with France. Governor Phillip decided that there should be a large bay set aside for neutral ships visiting the harbour.

New England *NSW* was known to the Aboriginals as Arrabald. The tableland was crossed by John Oxley and his party in 1818. The first settler was H.C. Sempill, who established a station named 'Walka' (Walcha) in 1832. After several years a number of squatters settled in the district and named it

New England, though part of the territory was for some years known as New Caledonia by the local residents.

New Holland. The name New Holland was frequently used for the continent until the early part of the nineteenth century, when it was displaced by Flinders' Australia, which see. The older name originated from Tasman, who called it Compagnis Nieu Nederland, which became anglicised as British interests increased. The name still survives in scientific nomenclature in the form *novae hollandiae*.

New Lambton *NSW* is a suburb of Newcastle which takes its name from a colliery established by J. and A. Brown in 1868. The former landowner, Dr Mitchell, a chief shareholder in the Lambton coal mine, let the possession of the land slip through his fingers by not making his payments good. When Mr A. Brown discovered this, he rushed to the government office and had just finished making his payment for the land and the mineral rights and was leaving the office when he met Mr Morehead, superintendent of the Lambton mine coming in. They said good morning in a gentlemanly manner as they passed. The Browns decided to call their colliery 'New Lambton' in order to trade on the established reputation of Lambton coal. This incensed R.A. Morehead, the Superintendent of the Scottish Australian Mining Company. The Lambton Company in their advertisements denied any connection with this new company and emphasised that the name, New Lambton, was used without their sanction. The bitterness which the naming engendered was never forgotten by the Lambton management.

New Norcia *WA* is the site of a monastery established by Benedictine monks from Spain in 1846. It was named after Norcia in Italy, the birthplace of Saint Benedict, by Dom Rosendo Salvado.

New Norfolk *T* is a town in the Derwent Valley about thirty-five kilometres from Hobart. It got its name because after the Norfolk Island colony closed in 1813, many former Norfolk Island inhabitants came to the area.

New South Wales The name New Wales was given to the whole of the eastern seaboard of Australia, i.e. present-day New South Wales and Queensland, by Captain James Cook in 1770, but he later altered the name to New South Wales. He gave no reason for his choice. There is a saying that the initials, which are now so well known, stand for Newcastle, Sydney, and Wollongong. There have been agitations for a change of name on a number of occasions. The name was conferred by Cook on 22 August 1770. Having landed on Possession Island, he wrote: 'Having satisfied myself of the great Probability of a passage, thro' which I intend going with the Ship, and therefore may land no more upon this Eastern coast of New Holland, and on the Western side I can make no new discovery, the honour of which belongs to the Dutch Navigators, but the Eastern Coast from the Lat of 38° S. down to this place, I am confident, was never seen or visited by an European before us; and notwithstanding I had in the Name of His Majesty taken possession of several places upon this Coast, I now once more hoisted English colours, and in the Name of His Majesty King George the Third, took possession of the whole Eastern coast from the above Latitude down to this place by the Name of New South Wales.'

New Town *T* was near the site of the government farm established at Cornelian Bay in the early days. Land was assigned to free settlers on New Town Bay (formerly Stainsforth Cove) just to the north; 'these settlements acquired the name of New Town in contradistinction of the original settlement of Hobart Town'.

Newcastle *NSW* was originally called King's Town after Governor Phillip King had the area surveyed in 1801 and established a small settlement at the mouth of the Hunter with convict labour to mine coal. It was soon abandoned. In 1804, King commissioned Lieutenant Charles Menzies to re-open the settlement. The name, Newcastle, after the place of that name in England, appeared for the first time in Menzies' commission. There were sixty-four people involved in the new settlement. Half were convicts and half soldiers. Because it was remote from Sydney, being sent as punishment to Coal River was dreaded by the convicts. The local pronunciation is New-KAR-sel rather than New-KASS-sel.

Newcomb *V* is a residential district east of Geelong City. It is named after Caroline Newcomb who, with Anne Drysdale, worked 'Boronggoo' station between the Barwon River and Point Henry from 1840 to 1845.

Newman, Mount *see* **Mount Newman**

Newport *NSW* is a suburb of Warringah Shire in Sydney's northern beaches area. Newport is one of the oldest settlements in the Manly-Warringah area. The first grant of land was made to John Farrell in 1843. For many years the only approach to Newport was by water and it was a port in fact as well as in name The Pittwater shore was a favourite picnic spot. In 1875 a Mr Jeanerett built a home at Newport and linked the spot to Manly by a line of coaches.

Newport *V* is a suburb of the City of Williamstown which is located across Phillip Bay from the City of Melbourne. Newport referred to the fact that a new port was built there.

Newtown *NSW,* a Sydney suburb, is attributed to one John Webster who opened a store called the New Town Store to distinguish it from stores in other localities. The first available reference to the locality as Newtown is from the *Sydney Gazette* of 1832 which said 'the neighbourhood about the spot of Devine's farm has obtained the name Newtown'. However, Newtown was not proclaimed a municipality until 12 December 1862.

Newtown *V* is a city in Greater Geelong. It became a borough in 1858. The town of Newtown and Chilwell dates from 1924. It became a city in 1949. It was renamed Newtown in 1967. The name occurs in Cornwall, England, but has the literal meaning of a new town.

Nhill *V* is a township seventy-four kilometres north-west of Horsham. The name comes from an Aboriginal word, possibly *nyell,* meaning 'place of spirits'.

Nhulunbuy *NT* is the Aboriginal name for a nearby hill called Mount Saunders by Matthew Flinders. The town of that name was founded by an aluminium company, which was formed in 1965 to mine the bauxite deposits of Gove Peninsula. The name is pronounced 'Nullunboy'.

Nielsen Park *NSW* is located on Sydney Harbour in the Municipality of Woollahra. Nielsen Park was named in honour of N.R.W. Nielsen who, as Minister of Lands, in 1910-1911 authorised the purchase of land along the foreshores there for use as a picnic reserve. The far-sighted Nielsen would no doubt have approved the fact that his park is now part of Sydney Harbour National Park.

Nimbin *NSW* Roland Robinson states that the Aboriginal, Alexander Vesper, informed him that Nymbun is the name of a spirit in the form of a small man who lives inside the mountain. The Nymbun would be the *barnyunbee,* the totem of the place, and the mountain would be the *jurraveel,* or totemic site.

Nimmitabel *NSW* An Aboriginal name meaning 'meeting place of many waters'. There can be few names that have taken so many forms over the years—Nimmitybelle, Nimitybell, Nimithyball, Nimity-Bell, Nimoitebool, Nimmittabel, etc.

Noarlunga *SA* is a City located approximately twenty-five kilometres south of Adelaide on the Fleurieu Peninsula.

Noarlunga is generally agreed to be derived from an Aboriginal word meaning 'fishing place', although one source states that it is derived from an Aboriginal word meaning 'the place on the hill'. *The Observer* on 13 April 1844 said Noarlunga was derived from *Nurlo-ngga; nurlo* meaning 'a curvature' or 'elbow', and *ngga* meaning 'on' or 'upon'. The town was laid out and named in 1840. It was sometimes called Horseshoe from the curve in the river.

Nobbys Head *NSW* Though sighted by Captain Cook, the island was first visited by Lieutenant-Colonel William Paterson, who called it Coal Island. In 1797 Lieut John Shortland named it Hacking's Island. Later it became known as Nobby's Island. The earliest mention of this name was in 1810. When connected to the mainland it became Nobby's Head. The Aboriginal name is Whibay-Garba.

Noble Park *V* is a suburb of the City of Springvale on the outskirts of Melbourne. Noble Park was named by Alan Frank Buckley after his son, Noble.

Nollamara *WA* in Stirling, a city in the Perth metropolitan area, is a local Aboriginal name for the West Australian wildflower known as the kangaroo paw. The name was suggested in 1954.

Noosa Heads *Q* is an Aboriginal name. It was first called Cape Bracefield, after James Bracefield, an escaped convict, who lived here together with James Davis among the Aboriginals and were known as the wild white men. Bracefield helped Mrs Fraser to reach civilisation and then returned to the tribe but six years later he was found by Andrew Petrie and pardoned.

Norfolk Island is a territory under Australian authority in the Pacific Ocean 1676 kilometres north-east of Sydney. It was discovered by James Cook on 10 October 1774. He took possession of the island conferring the name Norfolk Isle 'in honour of that noble family'.

Norfolk Plains *T* was an early name for the Longford area which was one of the earliest settlements in the north of the island. The Norfolk Islanders who gave it this name arrived in the early 1800s but do not appear to have travelled up the Lake River for any distance.

Norlane *V* is a suburb of Geelong in the Shire of Corio. It is named after Norman Lane, a World War II serviceman.

Normanhurst *NSW*, a suburb of Hornsby Shire, is named after Norman Selfe, an engineer who lived in the area.

Normanton *Q* takes its name from the Norman River which was named by J. L. Stokes after Captain Norman of the *Black Diamond.*

Norseman *WA* was a reef named by a gold prospector after one of his horses in 1893. Whilst returning from unsuccessful prospecting in Coolgardie, Laurie Sinclair stopped at Dundas and made his find of the Norseman reef some twenty-four kilometres north of the town. He named his find after his horse, Norseman, but he was also evidently proud of his Norse ancestry. The townsite was gazetted on 24 May 1895.

North Arm Cove *NSW* was the site of a proposed city planned by H. F. Halloran and Co. which has never taken shape.

North Rocks *NSW* is a locality in the Baulkham Hills Shire on the north-western outskirts of Sydney. It was once the outer area of Parramatta where sandstone was quarried. Many old buildings such as Parramatta Gaol were built of the stone from the area. It is said the area was referred to as North Rocks.

North Sydney *NSW* is a municipality on the northern side of Sydney Harbour Bridge. It is a name of fairly recent origin. The explorer and surveyor, Sir Thomas Mitchell, surveyed St Leonards in 1828 and for many years the whole district was known by that name. The change to North Sydney caused much controversy among early residents, but eventually a bare majority voted for the new name on the grounds of prestige and for commercial reasons and the name North Sydney was adopted.

North-West Cape *WA* was named by P.P. King on 10 February 1818. The cape had been sighted by several Dutch explorers in the 1600s. Nicolas Baudin named it Cape Murat on 22 July 1801 after a brother-in-law of Napoleon but the name was replaced by King's choice.

Northam *WA* was named by Governor Stirling after Northam in Devonshire. In October 1830 Ensign Robert Dale led a party of colonists over the Darling Range into the valley of the Avon. Stirling named Northam after Dale's expedition, the town-site being declared in 1833.

Northbridge *NSW is* a suburb of the Municipality of Willoughby on Sydney's North Shore. Northbridge gets its name from the old suspension bridge which connects the suburb with Cammeray and the other side of the gully. It was built by the North Sydney Tramway and Development Co. to open up the Northbridge area as a residential suburb, then known as Gordon's Estate. The bridge, of ornate design with castellated towers, was widely admired. It contained some 2000 tonnes of iron and steel and cost one hundred thousand pounds. It was begun in 1889 but the economic depression of the 1890s delayed the opening for some years. In 1912 the state government took over the bridge and ran a tram service to Northbridge, but this ceased in 1936 when the bridge was declared unsafe. In 1939 it was reopened as a reinforced concrete arch and girder bridge. It has now been classified by the National Trust as a historic asset.

Northcote *V* is an industrial city close to the centre of Melbourne. According to most history books, the treaty between John Batman and the Port Phillip Aborigines was signed in Northcote in 1835. But there is now some dispute as to whether Batman even visited Northcote. Northcote is named after Baron Henry Stafford Northcote, who served as Governor-General of Australia from 1904 to 1908. Northcote became a borough in 1883 and was proclaimed a town in 1890. It was proclaimed a city in 1914.

Northern Territory When the territory was annexed to South Australia on 1863, it was described as comprising 'all the country to the northward of the 26th parallel S latitude and between the 129th and 138ths degrees of E longitude'—and so received its name. Before this, it was part of New South Wales, and from 1911 has been administered by the Commonwealth. Early names for 'the Territory' were Alexandra Land and Prince Albert Land. During World War II, there was a proposal to name it Churchill Land and in 1954, when Queen Elizabeth II visited Australia, Elizabeth, but 'the Territory' has survived. Alexandra Land was suggested by J. McDouall Stuart in November 1864 in honour of Princess Alexandra, and was officially adopted for some time. Stuart's proposal was embodied in a paper sent to the Royal Geographical Society.

Northfield *SA* is a locality in the corporate City of Enfield in the Adelaide metropolitan area. Northfield is a descriptive name.

Northmead *NSW* is a locality in the Baulkham Hills Shire on the north-western outskirts of Sydney. Northmead was originally referred to as North Meadow.

Norwood *SA* is a suburb of the corporate city of Kensington and Norwood in the Adelaide metropolitan area. The oldest part of Norwood was laid out in 1838 by Samuel Reeves. A Lands Department document registered in 1849 shows where Reeves laid out section 260 as the village of Norwood.

Presumably he called it after Norwood in England, then a village a few kilometres south of London but now within the Greater London area. Some of the original streets of Norwood also occur in Norwood, England; for example Beulah and Sydenham. After the name, Norwood, was given to the four sections. the north-west part of the district which was settled in 1838 was known as Old Norwood.

Nowhere Else *T* is the name applied to a farming district about four kilometres west of Sheffield. It is supposed to have originated from locals so answering when asked where a certain dead-end road led to.

Nowra *NSW* is an Aboriginal word for 'black cockatoo'.

Nullarbor Plain WA comes from Latin words meaning no tree. *(Null* is Latin for 'no' and *arbor* is Latin for 'tree'). This descriptive name was applied by Alfred Delisser who examined it at greater length than any other explorer in 1865-66.

Nunawading *V* is a city in the Melbourne metropolitan area. The name comes from an Aboriginal word *numphawading* meaning 'ceremonial ground', or 'battle ground where people come together'.

Nundah *Q* is a locality in the Brisbane metropolitan area and was named by the Railways Department in 1884. It is an Aboriginal word meaning 'chain of waterholes'.

Nuriootpa *SA is* derived from an Aboriginal word meaning 'meeting place' or 'neck of the giant'. It was formerly an area for corroborees and a bartering point for several Aboriginal tribes. The town was subdivided by William Coulthard in 1850.

Nyngan *NSW* An Aboriginal term meaning 'crayfish', 'mussel', or 'many creeks'. T.L. Mitchell recorded the name in 1835 as Nyingan, and said it meant 'long pond of water'.

O.K. *Q* This deserted copper-mining township is said to have acquired its peculiar name from the miners when a jam tin of this brand was found beside the shaft.

Oak Park *V* is a suburb of the City of Broadmeadows in the Melbourne metropolitan area. Oak Park takes its name from a house built by John Pascoe Fawkner between about 1839 and 1853.

Oaklands *SA is* the name of a place on Yorke Peninsula christened in 1854 by R.D. Anderson (from an abundant growth of she-oaks).

Oaklands Park *SA* is the name of a locality in the corporate city of Marion in the southern part of the Adelaide metropolitan area. Oaklands Park takes its name from the Oaklands Estate founded by Samuel Kearne in 1844.

Oakleigh *V* is a suburban city about fifteen kilometres south-east of Melbourne. Monash University is located within the city's boundaries. Oakleigh was named by the Executive Council when it approved Surveyor Robert Hoddle's plan for the village in 1853. The reason for this name has never been determined. Oakleigh became a shire in 1879. It was declared a town in 1924 and was proclaimed a city in 1927.

Oatlands *T* was named in 1821 by Governor Macquarie because it reminded him of places where oats grew in his native Scotland.

Oatley *NSW*, a suburb of the Sydney municipalities of Kogarah and Hurstville, was a grant made to James Oatley in 1833 but not developed until after his death. James Oatley was an emancipist farmer and clockmaker who made the clock in the turret of the Hyde Park Prisoners' Barracks which tourists visit in Queens Square in Sydney. Governor Brisbane is said to have granted him the land as a reward for making the clock.

O'Connor *ACT* honours Richard Edward O'Connor (1851–1912), one of the founders of the Constitution.

O'Halloran Hill *SA* is a hill and a suburb of the corporate city of Marion in the southern part of the Adelaide Metropolitan area. O'Halloran Hill is named after one of the early settlers in the district, Major Thomas Shuldham O'Halloran, who took up land in 1839. He was appointed Commissioner of Police and a police magistrate by Governor Gawler in 1838 but retired in 1843. He was a member of both the nominee and elected legislative councils. He was also a governor of St Peters College. He died in 1870 at the age of 73.

Olga, Mount *see* **Mount Olga**

Olinda *V* is a holiday resort on Mount Dandenong, named for Alice Olinda Hodgkinson, daughter of Clement Hodgkinson, a prominent surveyor and public servant.

Olympic Dam *SA* The camp site took its name from the dam constructed and named by a lessee of Roxby Downs Pastoral Lease.

O'Malley *ACT* is named after King O'Malley (1854-1953) who, as Minister for Home Affairs between 1910 and 1916, arranged for a competition to design the national capital.

Omeo *V* is a town on the Mitta Mitta River in the Great Dividing Range in the far north-east of the state. Gold was discovered there in December 1851. The Aboriginal word *omeo* means 'mountains'.

One Tree Hill *SA* is a township in the Munno Para district on the outskirts of the Adelaide metropolitan area. The hill referred to in the name is about three kilometres from the present township and the solitary tree that was its distinctive feature was cut down in the 1890s. Five gums have since been planted to replace the original. The One Tree Hill Inn, opposite the site of the original one tree, was first licensed in 1851. It was used for many years as the meeting place of the District Council of Munno Para East.

One Tree Hill *V*, a locality of the Shire of Sherbrooke on the outskirts of Melbourne, owes its name to the activities of an unknown survey party who felled all the trees on the hilltop except for a giant ash sometime before 1861.

Onkaparinga *SA* is a district in the Mount Lofty Ranges in the Onkaparinga River catchment area. Onkaparinga is derived from an Aboriginal word. A number of possible derivations have been suggested. They include *ponkporringa,* 'shadows in the water'; *unkaparinga, ingangkiparri* and *ungkaparinga* meaning 'another river', 'plentiful'; *onkaga,* 'mother river'; *ngankiparri-unga,* 'place of the women's river'. The river was discovered by Captain Collett Barker in 1831 when exploring the gulf coastline, seeking the outlet from Lake Alexandrina.

Onslow *WA* is called after Sir Alexander Campbell Onslow, Chief Justice of Western Australia at varying times between 1874 and 1901.

Orange *NSW* was named in 1833 by the explorer Sir Thomas Mitchell who conceived the name as a tribute to the Prince of Orange, later King of the Netherlands. Mitchell and the Prince had been associated during the Peninsular Wars in Spain. Orange was proclaimed a village in 1846, a municipality in 1860 and a city in 1946.

Oraparinna *SA* is an Aboriginal word meaning 'a creek with tea-trees.'

Orbost *V* is a township by the Snowy River in East Gippsland. Its Gaelic name means 'winged island' and derives from the Isle of Skye, home of the uncle of squatter Archibald McLeod.

Ord River *WA* was discovered by Alexander Forrest on 25 July 1879 and named after Sir Harry St George Ord, Governor of Western Australia from November 1877 to April 1880.

O'Reillys *Q* is named after the pioneering family that owns the land and established the resort there. One of its members, Bernard O'Reilly (1903-1975), led a rescue party after an air crash in the rugged Macpherson Ranges in 1937. The story of the rescue and the O'Reilly family is told in *Green Mountains and Cullenbenbong* by Bernard O'Reilly, published in 1949.

Orelia *WA*, a residential suburb of the town of Kwinana near Perth, takes its name from a 390 tonne barque which arrived on 12 October 1829 with twelve passengers.

Ormond *V* is a suburb of the City of Moorabbin in the Melbourne metropolitan area. Ormond is named after Francis Ormond, a philanthropist who gave money to Melbourne University in the 1800s.

Osborne Park *WA* in Stirling, a city in the Perth metropolitan area, was named after William Osborne, a butcher who owned an abattoir and land along the Wanneroo Road and who was elected to the Road Board in 1875.

Ossa, Mount *see* **Mount Ossa**

Otway, Cape *see* **Cape Otway**

Ourimbah *NSW* is an Aboriginal word. It may mean 'bora' or 'ceremonial ground'. But like most derivations of Aboriginal words in the Sydney area, this is dubious.

Ouyen *V* is a township in the Mallee area 104 kilometres south of Mildura. The Aboriginal word means 'ghost waterhole' or 'wild duck'.

Ovens River *V,* a north-flowing tributary of the Murray River rising near Mount Hotham, was named by the explorer William Hovell in 1824 in honour of Major John Ovens, a soldier and explorer.

Oxley *ACT* recalls the surveyor, John Joseph William Molesworth Oxley (1785–1828), who explored northern New South Wales and southern Queensland between 1817 and 1823.

Oxley *Q* is a locality in the Brisbane metropolitan area and is named after Surveyor-General John Oxley (1785-1828) who explored the Brisbane River in 1823 and selected the site for the first convict settlement there.

Oyster Bay *NSW* in the Sutherland Shire derives its name from the abundance of oysters grown in the bay. Named before the Thomas Holt period, 1860-1883; it appeared on Surveyor Wells' map of 1840.

Paddington *NSW* is a suburb in the Municipality of Woollahra. It is one of Sydney's fashionable inner suburbs with many renovated terrace houses. Paddington was probably named after the London suburb by its first European residents who were granted land above Rushcutters Bay by Governor Bourke. The name, Paddington, was in colloquial use during the 1820s and in 1842, when an auctioneer, Mr Samuel Lyons, advertised the sale of land in this area in the *Sydney Morning Herald* (3 January and 17 September 1842), it was referred to as being 'a portion of the celebrated estate of Paddington'. Quite quickly the name came to be given to the whole neighbourhood. Upon the passing of the Municipalities Act of 1858, 172 local residents petitioned the Governor and in 1860 the Municipality of Paddington was incorporated.

Padstow *NSW* is a suburb of Bankstown, a city located in the western part of the Sydney metropolitan area. It was named after Padstow in Cornwall, England, which is a corruption of Petrockstow, where there is a church dedicated to St Petroc, a sixth century Welshman, educated in Ireland.

Page *ACT* is named after Sir Earle Page (1880-1961), who was Prime Minister in 1939.

Pagewood *NSW* is a suburb of the Sydney municipality of Botany. Pagewood was originally called Daceyville No. 2, to distinguish it from the Mascot Scheme. However, it was changed in 1929 to Pagewood in honour of F.J. Page, a highly respected member of the Botany Council.

Pakenham *V*, a town in West Gippsland fifty-seven kilometres south-east of Melbourne, was named after General Pakenham, who served in the Crimean War. It was formerly known as Longford, after the English M.P.

Palm Beach *NSW* is located on the tip of Barrenjoey Peninsula in Warringah Shire. Palm Beach gets its name from the cabbage tree palms which grow there. It was once called Cranky Alice Beach.

Palm Island *Q* was named by Captain Cook on 7 June 1770 because of the cabbage palms growing there.

Palmer River *Q* was discovered by William Hann in 1872 and named after Sir Arthur Hunter Palmer, Premier of Queensland.

Palmyra *WA* was chosen after a competition run by the Melville Road Board. The original Palmyra was a city located about 200 kilometres north-east of Damascus in Syria. It was built by King Solomon and reached the height of its wealth and splendour under King Odenathus (A.D. 255-67) and his wife and successor Queen Zenobia. Streets in the modern suburb such as Zenobia, Solomon and so on are associated with the ancestral city in the Middle East.

Pambula *NSW* is Aboriginal for 'two streams' .

Panania *NSW* is a suburb of Bankstown, a city located in the Western part of the Sydney metropolitan area. It is an Aboriginal word meaning 'the sun rising and shining on hills'.

Pandoras Pass *NSW* was named by Allan Cunningham in 1823 when he was attempting to find a practical route between Bathurst and Liverpool Plains. He had difficulty in negotiating the Warrumbungle Mountains, and gave this hopeful name to the pass he discovered. The pass was negotiated successfully by John Oxley and himself two years later with a happier result than when Pandora, a character in Greek mythology, opened her box.

Panton Hill *V* is a locality in the Shire of Eltham on the outskirts of Melbourne. Panton Hill, was originally known as Kingston. It was renamed after John Anderson Panton who was appointed police magistrate in the district in 1862.

Para Hills *SA* got its name originally from the Little Para River. The Para Plains which adjoined the river got their name from the river and when the first settler, John Goodall, obtained land in 1850 in the adjoining hills he naturally called his property 'Para Hills Farm'. *Para* is an Aboriginal word for 'a stream of water'.

Para River *SA Para* is an Aboriginal word for 'a stream of water'.

Parachilna SA is Aboriginal for 'river with steep banks and a stony bed' according to A.W. Reed in his *Aboriginal Words and Place Names*. N.B. Tindale, however, says it is derived from Patatjilna, 'a place of peppermint gum trees'.

Paracombe SA is a locality in the District Council of Gumeracha in the northeastern section of the Adelaide metropolitan area. Jacob Hagen gave the name 'Paracombe' to the 400 acres he acquired from his friend John Barton

Hack in 1840. There is a Parracombe in Devonshire, England, which Hagen may have had in mind as both he and Hack originated from County Devon. It may be, however, that the name refers to the Little Para River itself which ran through the property. In early directories the name Paracombe frequently occurs as two words, Para Combe, which may possibly suggest that the name has a hybrid derivation from the Aboriginal word *para* meaning 'creek' or 'river' and the English word *combe,* 'a narrow valley'. The first recorded subdivision referring to the name Paracombe was dated 3 November 1863 and was surveyed by William Weir.

Paradise *SA* is a suburb of the corporate city of Campbell-town in the Adelaide metropolitan area. Joseph Jud, one of the early settlers in Paradise, was born in Gloucester, England and reached South Australia on 4 December 1837. After living in Hindley Street, Adelaide, for a short time, he acquired land at Paradise for gardening purposes and became famous for the quality of the water-melons and sweet-melons he grew. On 11 December 1850, Joseph Jud applied for a licence to build a 'bar' in front of his cottage. According to the official local history by J.T. Leaney, Joseph Jud erected the new hotel and called it the Paradise Bridge Hotel, a name derived from a property called 'Paradise' near Tetbury in the Cotswolds of Gloucestershire, presumably owned by Joseph Jud. Evidence indicates that the name, Paradise, was being used as early as 1853 when Joseph Jud described himself as being of Little Paradise on the River Torrens. A licensed victualler, Joseph Jud is shown in Department of Lands records as being of the Paradise Bridge Inn on the River Torrens in 1854 and in the following year he leased property which was referred to in the abovementioned records as 'The Garden of Paradise' on the Torrens.

According to Lands Department records the south branch of the Main North-Eastern Road (lower North-East road) was opened in 1854. In a document dated 13 July 1854 reference is made to Joseph Jud's house known as the Paradise Bridge Hotel. The bridge was officially opened on 13 August 1857 and named McDonnell Bridge after the governor although the bridge is now known as the Paradise Bridge.

Paralowie *SA* was named after the large house and citrus orchard nearby. The suburb stands on a portion of the land which originally belonged to the 'Paralowie' estate. *Paralowie* is an Aboriginal word—*para* meaning 'river', and *lowie,* 'water'.

Parkdale *V* is a suburb of the City of Mordialloc in the

Melbourne metropolitan area. Parkdale was named by the Railways Department after Mr Parker, the original owner of the land surrounding the present railway station.

Parkes *NSW* is called after Sir Henry Parkes, Premier of New South Wales, 1872–83 and 1887-91. The town was formerly Bushman's, from the mine named Bushman's Lead. It was changed to Parkes in 1873.

Parklea *NSW,* a locality in the City of Blacktown on the western outskirts of Sydney, was named by the subdividers in the 1900s. There is a reference to St Aidan's Church of England, Parklea, in some 1920 records.

Parmelia *WA,* a rapidly developing housing area of the town of Kwinana near Perth, is named after the ship that brought Captain Stirling to Fremantle on 1 June 1829. It carried sixty-nine passengers.

Parramatta *NSW* is a city and the commercial centre of Sydney's western suburbs. Parramatta was originally named Rose Hill by Governor Phillip. However, in 1791 he changed it to Parramatta, the name by which the Aboriginals knew the locality. Much later in its history, Parramatta was said to mean 'place where eels lie down'. However, a more probable explanation is that the real meaning is 'head of the river' as reported by Mrs Macarthur.

Parsley Bay *NSW* is located on Sydney Harbour in the Municipality of Woollahra. Parsley Bay is a traditional name and was probably an identification used by the first exploring boat parties in 1788. It is reasonable to assume that they found some edible plant resembling parsley growing there.

Pascoe Vale *V* is a suburb of the City of Coburg in the Melbourne metropolitan area. Pascoe Vale was named after the farm which John Pascoe Fawkner, one of the co-founders of Melbourne, bought in 1839.

Patch, The *see* **The Patch**

Patricks River *NSW* The river was discovered on St Patrick's Day in 1818. The discoverer was James Meehan. T. L. Mitchell was an advocate for the retention of the Aboriginal name and wished to change the name of the river, but was unable to do so because the Aboriginal name could not be discovered.

Patterson *V* is a suburb of the City of Moorabbin in the Melbourne metropolitan area. Patterson was named after J.B. Patterson, Minister for Public Works.

Payneham *SA* is a city about six kilometres east of the centre of Adelaide. It is named after Samuel Payne, an Adelaide tavern keeper whose town property entitled him to a country section. He divided his land into lots in 1838 and advertised it 'to the working classes of Adelaide'.

Peachy Belt *SA* is a locality in the district of Munno Para north of Adelaide. The currently accepted version comes from H.C. Talbot who says it was so named from a large clump of quandongs (native peach). William Pearson says he saw them in 1848. The name was recorded as Peachy Belt in a survey dated 1849. The diagram shows the area to be almost completely covered with trees, possibly quandongs. Another possible derivation is that it was named after P. Peachey who arrived in South Australia in 1841 and became an overseer at 'Parnaroo' station when his uncle George Williams owned it with Phillip Levi about 1855.

Peakhurst *NSW*, a suburb of Hurstville, was suggested by a school inspector, Mr Huffer, when consideration was being given to the establishment of a public school in 1871. A Wesleyan Day School was started in 1862 in Peake's Chapel, and the residents applied for the school to be taken over by the Council for Education as a provisional public school. The Wesleyan Church was built on land provided by Mr John Robert Peake. The name proposed by the school inspector was Peakehurst, but the 'e' was lost in later correspondence. When the post office was opened in 1885, it was called Peakhurst.

Pearce *ACT* is called after Sir George Foster Pearce (1870–1952), a leading member of the Senate from 1901 to 1938.

Pearces Corner *NSW*, a locality in Hornsby Shire on the North Shore of Sydney, recalls the slab huts built by the Pearce family at the junction that bears their name. Aaron Pearce (1786–1849) was a convict sentenced to transportation for life. He arrived in Sydney in 1811 and became a timber getter at South Colah. His sons remained in the district and in 1894 still had homes at Pearces Corner.

Pedder, Lake *see* **Lake Pedder**

Peel Island *Q*, was named after one of the discoverers of Port Denison. The Aboriginal name was Jeerkooroora.

Peel River *NSW* was discovered in 1818 and named by John Oxley after Sir Robert Peel, the British prime minister.

Pelsaert Group *WA*. This group of islands in the Houtman Abrolhos is a reminder of an early tragedy. In June 1629 the Dutch vessel *Batavia,* commanded by Francois Pelsaert, was wrecked on a reef. The survivors took refuge on one of the islands while Pelsaert undertook the dangerous voyage to Batavia in an open boat. During his absence the supercargo, Jerome Cornelius, led a mutiny during which 125 of the castaways were murdered. The group was named by J. Lort Stokes RN on a survey under Commander J.C. Wickham RN on the HMS *Beagle* in 1840.

Pemberton *WA* was named after Pemberton Walcott, son of an early settler at the Warren River. Walcott arrived in this area to settle in 1862. The area had become known as Big Brook, but when it became necessary to gazette a townsite it was felt a more distinctive name was required. The name Walcott was first suggested but was discarded by the Post Office, and Mr W.L. Brockman then suggested Pemberton. The town-site was gazetted 30 October 1925.

Pendle Hill *NSW* is a suburb in the Parramatta area. It was named after Pendleton in Lancashire, England at the request of George Bond who established a cotton-spinning mill in the area prior to the station being built.

Penfield *SA* is a suburb of the District Council of Munno Para. It was named by William Penfold, who was granted land in the district in 1850. In 1853, he opened the Plough and Harrow Hotel, and in 1856 he subdivided his section into allotments for a township, which he called Penfield.

Penguin *T* took its name from the fairy penguin. The name was bestowed in 1821 by Ronald Campbell Gunn, a distinguished botanist.

Pennant Hills *NSW,* a locality in Hornsby Shire on the North Shore of Sydney does not have a clear origin. One theory is that the district took its name from a signalling post stationed on the hill now known as Mount Wilberforce, from which both Sydney and Parramatta, the governor's place of residence, could be seen. However, no evidence can be found for this as the signalling posts were at One Tree Hill at Ermington and on May's Hill near old Government House at Parramatta. Another more likely suggestion is that it was named in honour of Thomas Pennant (1726–1798), a noted English naturalist and friend both of Sir Joseph Banks and of the father of Francis Grose, Lieutenant-Governor of New South Wales.

Penneshaw *SA* is a settlement on Kangaroo Island which was first called Hog Bay (the bay on which it stands, so named because of the pigs that were brought here from Tasmania). Governor Jervois was not enamoured of the name and changed it to Penneshaw, probably after his secretary, Dr F.W. Pennefather, and a Miss Shaw who had been visiting his family.

According to Rodney Cockburn's *What's In A Name*, Miss Shaw was the second daughter of Major-General George Shaw and was formerly head of the colonial department of the *Times* before becoming Lady Lugard, wife of the Governor and Commander-in-Chief of Hong Kong.

Pennington *SA* is a suburb within the corporate city of Woodville in the Adelaide metropolitan area. The land was subdivided in 1909 by Captain Alfred Hodgeman. According to the local council history, Hodgeman called the subdivision after his wife whose maiden name was Helen Pennington. The street names Alfred, Helen, and Hodgeman Road were named by Hodgeman. But according to Rodney Cockburn, the name (which occurs in three English counties) dates back to 1840.

Pennydale *V* is a suburb of the City of Moorabbin in the Melbourne metropolitan area. Pennydale is named after Councillor E.T. Penny who was Shire President from 1898 to 1900.

Penola *SA* is an Aboriginal word and is a corruption of Penaoorla, the Aboriginal name of the big swamp several kilometres from the township, and the name of the station founded in the early 1840s by Alexander Cameron, who founded the town. The government town was surveyed in 1867 and named Penola North. It was altered to Penola in the *Gazette* dated 20 February 1941 to agree with the railway station and private town.

Penrith *NSW* is a city about fifty kilometres to the west of Sydney. Penrith was named after Penrith in Cumbria, England where the town is situated on the banks of a river with hills behind. The name means 'chief ford'. Governor Phillip initially named the Australian settlement Evan after Evan Nepean. Governor Macquarie changed the name to Penrith in 1818.

Pentridge *V* is a locality in the City of Coburg in the Melbourne metropolitan area. Pentridge was named after Thomas Pentridge, an Irish settler, who had his home by Merri Creek in the early days of the district. Pentridge Stockade (later Pentridge Jail) was established in 1851.

Peppermint Grove *WA* is a shire in the Perth metropolitan area. The shire trees was originally heavily wooded, and among the species of trees that grew there were the peppermint trees that gave the shire its name.

Perisher Valley *NSW*, a ski resort, is said to have got its name because of the fact that the valley is so cold that the early settlers felt people were likely to perish there in winter.

Perry Lakes *WA* commemorates Mr H. Perry, a dairyman from whom the Perth Council purchased the area of land originally known as Limekilns Estate in 1917.

Perth *WA* was named by Governor Stirling in honour of Sir George Murray, the Colonial Secretary, who was born in Perth, Scotland and at that time represented Perth Shire in the House of Commons. Perth was founded on 12 August 1829.

Perth Street Names. The first accurate plan of Perth street names was prepared by Alfred Hillman under the direction of the Surveyor-General and issued in 1838. Many of the names have been changed since the original plan was drawn up.

Adelaide Terrace was not named until 1830. It was named after Queen Adelaide, the wife of William IV; Barrack Street was so called because the soldiers who came out to protect the colonists were housed in the locality of the Treasury Buildings, first in tents and later in permanent barracks; Hay Street was named after Robert William Hay who was one of the permanent undersecretaries for the colonies at the time of Perth's foundation; Murray Street was called after Sir George Murray, Secretary of State for the colonies; Wellington Street is named in honour of the Duke of Wellington, who won the Battle of Waterloo and was first Lord of the Treasury in Britain during the period of foundation of the colony in 1823; William Street honours King William IV (first known as King William Street); St Georges Terrace is named after St George, the patron saint of England.

Peterborough *SA* The town was formed in 1880 and named Petersburg in honour of Peter Doecke, one of the early settlers who took up land in the district. In 1917 when anti-German feeling was at its height, there was a proposal to substitute the Aboriginal name Nullya or Nelia, but compromise resulted in Petersburg being anglicised to Peterborough and gazetted on 10 January 1918.

Petermann Range *NT* was discovered by Ernest Giles in 1874 and named after Dr Augustus Petermann, a German geographer who had taken an interest in Australian exploration.

Petersham *NSW,* a suburb of Marrickville, was named by Major Francis Grose after his native village of Petersham in Surrey.

Petrie *Q* is a locality in the Brisbane metropolitan area and honours a pioneer, Andrew Petrie (1798–1872), a builder and architect, who was one of the first free settlers in Brisbane. He came to Moreton Bay as clerk of works in 1837 and when the convict colony closed in 1839 he stayed on as a free settler. He designed and built many of the important buildings in the settlement. He also explored the Glasshouse Mountains and the Mary River.

Pewsey Peak; Pewsey Vale *SA* Named by Joseph Gilbert after his home town in Wiltshire. Before this he used the Aboriginal name, Karrawatta, but changed it because of confusion with 'Tarrawatta', John H. Angas's estate.

Phillip *ACT is* named in honour of Captain Arthur Phillip (1738–1814), the first governor of New South Wales, from 1788 to 1793.

Phillip Island *V* in Western Port was first discovered in 1798 by George Bass, who described it as 'a high cape, like a snapper's head', and again by Lieutenant Grant in 1801 in the *Lady Nelson* who also noted the resemblance, hence the early name Snapper Island. After a short period when it was called Grant Island, after Lieutenant James Grant, it was named after Governor Phillip. Baudin's name, Ile des Anglais, did not survive.

Phillip, Port *see* **Port Phillip**

Phillipson, Lake *see* **Lake Phillipson**

Pialba *Q* is an Aboriginal word meaning 'small green crabs with red claws'.

Pialligo *ACT* is an Aboriginal place name associated with the foundation of settlement in Canberra, that first appeared on Surveyor Robert Dixon's map of 1829.

Pichi Richi Pass *SA* is derived from an Aboriginal name for a bush called pichiri in South Australia and piturie in Queensland. The Aborigines used to chew the leaves and bark of the plant, which contain a powerful narcotic. The pass was discovered in 1843 by William Pinkerton, of Pinkerton Plains near Quorn. It was first known as Richman's Pass after John Richman, an early settler, who held pastoral country there.

Pickering Point *NSW*, a locality in Manly, is a headland of North Seaforth, and was frequently used as a camp site by sailing enthusiasts, William Harry and Fred Pickering.

Picton *NSW* was named by Major Antill after General Sir Thomas Picton, a British general who served in the Peninsular War and was killed at Waterloo. Governor Macquarie had called it Stonequarry, the village being on

Stonequarry Creek. The new name came into use in the early 1820s.

Pieman River *T* is named after a notorious character 'Jimmy the Pieman' who escaped three times from the penal settlement at Macquarie Harbour. On his first escape he was captured near the mouth of the river named after him. The Aboriginal name was Corinna.

Pilliga *NSW* is Aboriginal for 'place of swamp oaks' or from *biligha,* meaning 'head of scrub oak'.

Pinchgut *NSW* in Sydney Harbour is the popular name for the island on which Fort Denison stands. It probably comes from a nautical term meaning a narrow channel although the name is normally explained by a much more picturesque story about convicts being imprisoned on the island with only bread and water rations. In 1797, Governor Hunter decided to make an example of a murderer named Morgan by ordering him to be hanged in chains on a gibbet erected on the pinnacle of Pinchgut Island. The skeleton was allowed to remain there for three years. The fort on Pinchgut was named after Governor Denison.

Pindimar *NSW* is an Aboriginal term meaning 'black possum'. It was the site in the past of a state-run fishery, of a shark fishing plant and of an ambitious city proposal plan.

Pine Creek *NT* The locality was discovered during the survey for the Overland Telegraph Line and was named on account of the many pine trees on the bank.

Pinjarra *WA* is a name of Aboriginal origin but the meaning is not known. In the 1830s the settlement was known as Pinjarrup.

Pinkenba *Q* is a locality in the Brisbane metropolitan area and is an Aboriginal word meaning 'place of land tortoises'.

Pioneer River *Q* was originally called the Mackay after Captain John Mackay who found the Valley in 1860 and returned in 1862 to establish the initial settlement of Greenmount. The river was later renamed to honour a visit by HMS *Pioneer* under the command of Commodore Burnett.

Piper, Point *see* **Point Piper**

Pirie, Port *see* **Port Pirie**

Pitt Town *NSW* was named by Macquarie after William Pitt, the British prime minister who founded New South Wales.

Pittwater *NSW* was named Governor Phillip on 5 April 1788 after the Prime Minister of England, Pitt the Younger.

Pleasant, Mount *see* **Mount Pleasant**

Plenty *V* is a locality in the Shire of Diamond Valley on the outskirts of Melbourne near the Plenty River. Plenty was named by Joseph Gellibrand, a companion of John Batman, the founder of Melbourne. Gellibrand foresaw the district would be a land of plenty.

Plumpton *NSW,* a locality in the City of Blacktown on the western outskirts of Sydney, recalls Walter Lamb (1825–1906) who established the cannery and fruit preserving works on his Woodstock estate and was a promoter of coursing. He introduced the new Plumpton system of coursing to the colonies at his estate and later sold the grounds to the New South Wales Coursing Club. The area became known as Plumpton.

Poatina *T* is an Aboriginal word meaning 'cavern' .

Poeppel Corner *Q SA NT* Augustus Poeppel was the South Australian Government Surveyor who was sent in 1879 to survey the border between Queensland and South Australia along the 141st meridian from Cooper Creek northwards and along the 26th parallel of latitude to the 138th meridian, which forms the boundary between Queensland and the Northern Territory. The hardwood peg he drove into the ground to mark the junction of South Australia, Queensland, and Northern Territory became known as Poeppel's Peg. Further surveys were carried out by W.H. Cornish who discovered that Poeppel had placed the post 15 chains too far to the west, owing to an error in his chain. The "peg" was therefore shifted to the new position. It has since been replaced by a concrete cylinder, but the traditional ceremony of hanging one's hat on the Poeppel Peg is still observed by the occasional passer-by at Poeppel Corner.

Point Danger *NSW V* was named by Captain Cook on 16 May 1770. As there were breakers on the shoals, he wrote in his Journal: 'The point of which these shoals lay I have named Point Danger.'

Point Dinosaur *see* **Dinosaur Point**

Point Henry *V* is a low headland which forms the eastern arm of Corio Bay. It is named after Captain Edwin Whiting's brig *Henry* which anchored off the point on 16 June 1836. The point was marked on Whiting's survey map of Geelong Harbour, November 1836.

Point Hicks *V* was probably the point sighted by Captain Cook's men on 20 April 1770. Zachary Hicks, a lieutenant on Cook's *Endeavour,* was the first British seaman to sight

Australia's eastern mainland coast. The point cannot be determined for certain, but historians believe it was the promontory later known as Cape Everard. The name was changed to Point Hicks in 1970.

Point Lucinda *see* **Lucinda**

Point Macleay *see* **Macleay, Point**

Point Pickering *see* **Pickering Point**

Point Piper *NSW* is located on Sydney Harbour in the Municipality of Woollahra. Point Piper commemorates Captain John Piper, a military officer who became one of the richest men in Sydney as official collector of customs and harbour dues. He built a mansion at Point Piper and entertained so extravagantly that he earned the title 'prince of Australia'. Governor Darling, who arrived in 1825, dismissed him for neglect of his duties. Deprived of his lucrative post, Piper was soon in serious financial trouble. He tried to commit suicide by jumping out of his boat but was rescued by his boatmen. He died a poor man at Bathurst in 1851 at the age of seventy-eight.

Point Solander *see* **Cape Solander**

Point Sutherland *NSW* in the Sutherland Shire was named by Captain Cook in memory of Forby Sutherland, a sailor on the *Endeavour*, who died at Botany Bay and who is buried at Kurnell.

Pooraka *SA* is an Aboriginal name said to mean 'dry creek'. The district was formerly known as Montague Farm, after Sir Montague L. Chapman, the owner of Killua Castle, County Westmeath, Ireland. The post office was renamed Pooraka in 1916.

Port Adelaide *SA* is a city and harbour thirteen kilometres north-west of Adelaide. The Aborigines of the Kaurna tribe called the Port River, Yertabulti, meaning 'salt swamp that grew nothing'.

The site of the original port, developed inside the Port River, proved to be difficult. The mud, the mangroves, the shallowness of the water and the high tides all caused problems, and the place soon became known as 'Port Misery'. A new port was opened in 1840.

Port Arthur *T* was founded as a penal settlement in 1830 by Governor Arthur.

Port Augusta *SA* was named after Lady Augusta Young, the wife of the Governor, Sir Henry Fox-Young, in 1852. The harbour was discovered by John Grainger and A.L. Elder in the *Yatala* on 24 May 1852. They wrote a letter to the Colonial Secretary telling of their discovery of 'the new Port (which we have taken the liberty of calling Port Augusta)...' The Aboriginal name was Kurdnatta, 'place of drifting sand'.

Port Broughton *SA* is a small shipping port on the west coast of Yorke Peninsula, about 170 kilometres north-west of Adelaide. It was surveyed in October 1871 by Captain Henry David Dale of the schooner *Triumph,* the first vessel to visit there. The port is not on the River Broughton but it is the outlet for the agricultural area and town officially designated Broughton in 1869 but better known as Redhill because of the large, red hill nearby. It was altered to Broughton in a gazette of 1840. The River Broughton was named by E.J. Eyre in May 1834 after William Grant Broughton, the first Bishop of Australia. The town of Broughton took its name from the river and although the town is thirty kilometres inland from the port, it gave its name to the port.

Port Campbell *V,* a coastal township sixty-six kilometres south-east of Warrnambool, is named after Captain Alexander Campbell, a Scotsman who was in charge of the Port Fairy Whaling Station in the 1830s.

Port Curtis Q was named by Matthew Flinders in 1802. He called it after Admiral Sir Roger Curtis.

Port Elliot *SA* is a town on Horseshoe Bay, a small inlet on Encounter Bay, about eighty kilometres south of Adelaide. Governor Sir Henry Fox Young proclaimed the town a port in 1851. He named it after a friend, Sir Charles Elliot, who was then governor of Bermuda. It was named in 1850 and declared a port on 28 August 1851. With the building of the tramway from Goolwa to Port Elliot, it was expected to become the deep sea outlet for the riverboat trade to Goolwa.

Port Fairy *V,* a coastal township twenty-eight kilometres west of Warrnambool, was named because a sealer, Captain James Wishart, sheltered his ship the *Fairy* there in 1827 during a storm. In 1835, the locality was called Belfast when James Atkinson of Sydney took up land in the vicinity, but eventually it reverted to its original name. The Aboriginal name was Pyipgil.

Port Hedland *WA* was named after Captain Peter Hedland of the cutter *Mystery* in 1863 by a Mr Ridley who in company with a Mr Padbury and a second cutter *Tien Tsin* were examining country in the vicinity. Captain Hedland was later murdered by his Aboriginal crew. It was gazetted as a town-site on 23 October 1896.

Port Jackson *NSW* is the official name for Sydney Harbour. It was named after Sir George Jackson (1725–1822), Judge Advocate of the Fleet, by Captain Cook on 6 May 1770. Jackson was an early patron of the Yorkshire-born James Cook and followed his career with interest.

Port Kembla *NSW* comes from an Aboriginal word meaning 'abundant game'.

Port Latta *T,* was chosen in 1965 by the Nomenclature Board for the town and port site established in connection with the pelletising plant for the iron ore industry on the Savage River. *Latta* is a Tasmanian Aboriginal word meaning 'iron ore'.

Port Lincoln *SA* was named by the explorer Matthew Flinders who found the harbour in 1802. He called it 'Port Lincoln in honour of my native province'. Lincoln is a hybrid of Celtic and Latin and signifies 'a deep pool'. Nearby islands are named after eight of Flinders' men, who drowned near Cape Catastrophe. Thistle Island was named after Captain John Thistle, Flinders' sailing master, who led the ill-fated expedition in search of water. In 1841 an obelisk was erected by Lady Franklin on behalf of her husband, Sir John Franklin (leader of the ill-fated expedition to the north west passage in 1843), to the memory of Flinders under whom her husband had served as midshipman in the *Investigator.* Sir John was afterwards Governor of Tasmania. The Aboriginals knew the place as Kallinyalla. Baudin christened the place Port Champagny after the French Minister for the Interior. The town was surveyed in 1840.

Port Macdonnell *SA* was surveyed in 1860 and named after the then governor of South Australia, Sir Richard Graves Macdonnell.

Port Macquarie *NSW* is named after Governor Macquarie. It began as a penal settlement for notorious convicts in 1821. The settlement became a convict hospital in 1830 and was abandoned in 1847. Port Macquarie was proclaimed a municipality in 1887. The Shire of Hastings and the Municipality of Port Macquarie were amalgamated in 1981.

Port Melbourne *V* is a city located on Hobsons Bay about four kilometres south of the Melbourne General Post Office. The city was developed to serve the shipping needs of the port but now houses many industries including the main plants of the Australian aircraft building industry. In 1839 the surveyor, William Wedge Dark, called the area Sandridge after the ridge of sand dunes running along the beach and the district was known by this name until the name was changed to Port

Melbourne in 1884. Sandridge was created a borough in 1860. Port Melbourne became a city in 1919.

Port Phillip *V* Named after Captain Arthur Phillip, the first governor of New South Wales. Lieutenant John Murray sailed into the harbour in the *Lady Nelson* on 14 February 1802 and anchored near the present site of Sorrento. After a careful survey, during which some of the crew were involved in a skirmish with the Aboriginals, Murray had the British flag hoisted on 8 March and took formal possession of the territory, naming it Port King after the Governor, Phillip Gidley King. Governor King subsequently changed it to Port Phillip in honour of the first governor. Matthew Flinders arrived at the harbour in the *Investigator* in April, and wrote: 'This place, as I afterwards learned at Port Jackson, had been discovered ten weeks before by Lieutenant John Murray, who had succeeded Captain Grant in command of the *Lady Nelson*.' Until the separation of Victoria from New South Wales in 1851 the colony was known as the Port Phillip district.

Port Pirie *SA* is situated on the Port Pirie River and was surveyed in January 1872 and became the state's first provincial corporate city in 1953 although it became a corporate township in 1870. The Port Pirie River was named after the schooner *John Pirie* under Captain Thompson and was the first vessel to navigate the creek, which was known to the Aboriginals as Tarparrie meaning 'muddy creek', in 1845. The river was first shown on maps before 1846 as Samuel Creek after Samuel Germein, master of the government schooner *Waterwitch,* who first discovered the creek.

The vessel *John Pirie* was the third ship to arrive at Kangaroo Island under charter to the S.A. Company from 1835 . It was originally owned by John Pirie, a member of the original board of directors of the S.A. Company, who later became Mayor of London after having left home as a poor lad and eventually rising to be one of the largest shipbrokers in London; his story paralleled that of the fabled Dick Whittington.

Port Stephens *NSW* was named by Captain Cook in 1770 after an Admiralty official called Philip Stephens.

Port Victoria *SA* could have been one of the oldest towns in South Australia, if plans made in 1839 to establish the town had materialised. Nearly forty years elapsed before a further effort was made to establish the centre. In 1876 the town was named Wauraltee by G.E. Strangways. Confusion over the town's name continued until 1940 when it was officially named Port Victoria. It became a major wheat port between

1879–1949 and large grain clippers used the port until 1949. A group from the Adelaide Survey Association landed there in 1839, Messrs. Hughes, Cock, George James and others, and named the port after the schooner in which they sailed, the *Victoria*. Port Victoria was proclaimed a port in the *Government Gazette* dated 21 November 1878.

Portland *V*, a coastal town in the far south-west of the state, was named by Lieutenant James Grant after the Duke of Portland when he sailed past the bay in the *Lady Nelson* in 1800.

Portsea *V*, a bayside resort on Mornington Peninsula, was named after part of the town of Portsmouth, England.

Possession Island *Q* is a small, hilly, grass-covered island situated just west of Cape York on the far north-eastern coast of Australia. Captain James Cook landed there after sailing up the eastern coast of Australia. On 2 August 1770, in the name of King George III of England, he took possession of the entire eastern coast of the continent from latitude 38°S to latitude 10°30'S. Cook named this area New South Wales, and called the little island Possession Island. The ceremony of taking possession was commemorated in 1925 by the erection of an obelisk.

Powder Hulks Bay *NSW*, a locality in Manly, got its name from the hulks full of gunpowder that were once moored there. The explosives lighter, *Pride of England*, was acquired by the Explosives Department of New South Wales in 1882 for use in Middle Harbour and was anchored in this bay. The powder hulk *Behring*, built in the United States in 1850, was acquired in 1878. It was dismantled and sold in 1919. In 1925 the hull was towed to Sailors Bay where it was set on fire and burnt.

Prahran *V* is a densely settled city separated by the Yarra River from the centre of Melbourne. Prahran was recorded as Purraran, meaning 'land partly surrounded by water' by George Langhorne who established a mission to the Aboriginals in the area. The Surveyor-General, Robert Hoddle, spelt it Prahran and that is how it appeared on a map of Port Phillip in 1840, the year that the first crown land sales took place. Prahran was constituted a municipal district in 1855, a borough in 1863 and a town in 1870. It became a city in 1879.

Prairiewood *NSW* is a suburb of Fairfield on the south-western outskirts of the Sydney metropolitan area. It was the original name of the subdivision and the council decided to retain it on promulgation of the district.

Preston *V* is a city about ten kilometres north of Melbourne. Preston takes its name from Preston in England. A party of fifty settlers from Sussex arrived in 1850. They met in Woods store in High Street and gave the township its name because the countryside seemed to them to resemble that of the district around Preston in Sussex. The original name of the district was Jika Shire when it was created in 1871. The name changed to Preston in 1885. Preston became a town in 1922 and a city in 1926.

Prestons *NSW*, a suburb of Liverpool, is named after a local family who at the turn of the century managed the post office which was on the corner of Bringelly and Ash Roads.

Pritchard, Mount *see* **Mount Pritchard**

Proserpine *Q* is a town 1080 kilometres north of Brisbane on the Proserpine River. The explorer George Dalrymple named Proserpine which comes from Prosperina, the Roman name of the Greek goddess of fertility, Persephone.

Prospect *NSW*, a suburb of the Municipality of Holroyd and the City of Blacktown, in Sydney's western suburbs, was originally named Bellevue by Captain Arthur Phillip in 1790. He later changed it to Tench's Prospect Hill. It was from this point that he first saw the Blue Mountains.

Prospect *SA* is an inner suburban city located just north of the centre of Adelaide. In 1938, C. Johnson, solicitor and law agent, advertised land in Prospect Village. The name was apparently chosen 'because of the beautiful prospect the locality presented, well timbered with waving gums and shady trees'.

Puckapunyal *V* is a military camp and rural district sixteen kilometres west of Seymour. The Aboriginal name of the tribal group that lived by Mollison Creek was descended from Panyule (hill) tribe. *Pucka* meant 'to die', but the expert on Aboriginal names, Aldo Massola, explains the whole word as 'the middle hill'.

Pulpit Point *NSW* is a locality in the Sydney municipality of Hunters Hill. It was named by the pioneers from a rock formation at its tip. Races in the Hunters Hill regatta, the biggest annual aquatic event in the harbour in the 1860s, were to start here.

Punchbowl *NSW* is a suburb of the Sydney municipalities of Canterbury and Bankstown. The original Punch Bowl was on Cooks River where Punchbowl Road joins Georges River Road. A natural basin shaped like a punch bowl gave its name to a farm there early last century. When the railway to

Bankstown was opened in 1909, the station was named Punchbowl presumably because it was at Punchbowl Road.

Pymble *NSW*, a suburb of Ku-ring-gai municipality, is named after Robert Pymble (1788–1861), a silk weaver who came out from Herefordshire in 1821 and received a grant of land in 1823 for capturing a bushranger. His homestead at Pymble was built by convicts. Some of his descendants still live in the district.

Pyrmont *NSW* is a locality in the City of Sydney. Pyrmont was a favourite place for picnics in the early 1800s, and a young lady playfully christened it after Bad Pyrmont in Germany, which was a fashionable watering place. The Aboriginals had called it Pirrama.

Quakers Hill *NSW*, a locality in the City of Blacktown on the western outskirts of Sydney, was originally known as Douglas' Siding. The nearby hill had been named Quakers Hill after one of the early Quaker settlers. When in the early twentieth century, Rickards and Co. subdivided the land around Douglas Siding they named the estate Quakers Hill after the nearby hill.

Quarantine Head *NSW*, a locality in Manly, gets its name from the Quarantine Station established by Governor Bourke in 1832 and houses a collection of beautifully preserved buildings from the turn of the century. The first official use of the station was in 1837 when 56 out of 389 passengers died of disease on the way out to Sydney. The station has a number of engravings carved on the cliff by people who have spent time there. The earliest dates back to 1857 when the crew of the *Mariposa* left their mark there. The latest in a long line of engravings were left by the victims of Darwin's Cyclone Tracy after they were given accommodation at the station when their homes were destroyed on Christmas Day 1974.

Que River *T* is an upper feeder of the Huskisson River, which in turn flows into the Pieman. It is about eighteen kilometres north of the mining town of Rosebery. The origin of the name is unclear. The name Queva River was applied by V.D.L. Company Surveyor, Henry Hellyer in 1828, but to a different stream in the same area; but James Sprent's map of 1859 shows Que River on this stream. Whether this abbreviation was accidental or intentional is not known, nor

237

is the reason for Que River being placed in its present location on Charles Sprent's map twenty years later.

Queanbeyan *NSW* was first settled in 1828 when a holding was recorded under the name 'Queen Bean'. In 1838, the town was gazetted under its present spelling.

Queenscliff *NSW,* a suburb of Manly, was named during the reign of Queen Victoria 1837–1901 and appears in this form on old maps and plans of the locality.

Queenscliff *V* is a bayside port on the Bellarine Peninsula, thirty-one kilometres south-east of Geelong. It dates back to 1836 when it began as a fishing village consisting of a few wattle-and-thatch huts. It was originally known as Shortland's Bluff, but was later renamed in honour of Queen Victoria.

Queensland was named in honour of Queen Victoria at her own request. Before its separation from New South Wales, it was usually known as the Moreton Bay district. At the time of separation, several names were suggested—Cooksland, Flinders and Flindersland. The new colony of Queensland was officially created on 6 June 1859. This is now celebrated as Queensland Day.

Queenstown *T* was named after Queen Victoria. In 1896, the little shanty town of Penghana that had sprung up around the local copper smelters was destroyed by fire. The refugees moved to a newly planned town on the banks of the Queen River, and this settlement became known as Queenstown.

Quirindi *NSW* An Aboriginal term but the meaning is not clear. It is thought to be a corruption of *giyer warinda* or *giyer warindi,* variously rendered as 'dead wood', 'dead trees on a mountain top', and 'waters fall together'. The following variations have been recorded at different periods: Cuerindi, Cuerindie, Kuwerhindi.

Quorn *SA* is an abbreviation of Quorndon, a town in Leicestershire. The name was chosen by Governor Jervois as a tribute to his Leicester-born private secretary, J. Warner. The town was surveyed in April 1878, and proclaimed on 16 May 1878.

R

Raby *NSW* is a locality near Campbelltown on the outskirts of the Sydney metropolitan area. Raby was named by Alexander

Riley after his mother, whose maiden name was **Margaret Raby**. In the 1820s Riley imported the first Saxon merinos and established a stud at Raby. These stud sheep played an important part in the development of the Saxon merino flocks in the colony over the following thirty years. Raby was gazetted in 1976.

Radium Hill *SA* was the site of uranium deposits in South Australia about 100 kilometres south-west of Broken Hill. Uranium was first discovered there in 1906. It was mined for its radium during the 1920s and again during the 1950s. The mine closed in 1961. Radium Ridge was named by W.B. Greenwood in 1910 after the original uranium discovery located in the north Flinders Ranges.

Railton *T* was first shown on Roll Plan DII dated 1868. It is assumed that it derived from the then recently constructed line of the Mersey and Deloraine Tramway Company.

Rainbow Beach; Rainbow Reach *Q* Rainbow Reach is named after HMS *Rainbow* which anchored here in 1827. By a curious coincidence there was an Aboriginal legend about a man who fell in love with the rainbow that formed over Noosa Heads every evening.

Raine Island; Raine Passage *Q* Named after the discoverer, Thomas Raine, a seaman who became a pastoralist. In 1815 he set sail for China in the *Surry*, and explored parts of the Great Barrier Reef.

Ramsgate *NSW*, a suburb of the Municipalities of Rockdale and Kogorah, comes from the English seaside resort and is the name of the original subdivision.

Randwick *NSW* is a municipality in the Sydney metropolitan area. Randwick comes from an English village of that name near Stroud in Gloucestershire. Simeon Henry Pearce (1821–1886), the first mayor of Randwick, was born near the English village and was responsible for changing the name of the Australian suburb from Coogee as the area was then known. Pearce also transplanted the name High Cross to his new home.

Rankins Springs *NSW* is named after J.G.R. and A. Rankin, who had four selections in the district prior to 1866.

Rapid Bay SA was named by Colonel William Light after his vessel the *Rapid* when he visited the bay in 1836. Collet Barker probably visited the bay in 1831 before his crossing of the Murray Mouth and subsequent death. The Aboriginal name was Yatagolanaa.

Ravens Point *NSW*, a locality in the Municipality of Ryde, recalls William Raven, a captain and part-owner of the ship *Brittannia*. Buffalo Creek above Raven's old grant probably commemorates HMS *Buffalo*, on which Captain Raven had sailed.

Raymond Terrace *NSW* is named after Midshipman Raymond, who, in 1797, was sent by Lieutenant Shortland by boat up the Hunter River and remarked on the 'terraced' appearance of the trees at the junction of the Hunter and William Rivers. The locality was known for some time as Raymond's Terraces. Raymond Terrace was already known by its present name when Governor Macquarie visited it in 1812. The village was gazetted in 1837.

Reabold Hill *WA* incorporates the name of Frank Rea who was Mayor of Perth when the council bought the land in 1917.

Recherche Archipelago *see* **Archipelago of the Recherche**

Red Cliffs *V* is a soldier settlement area and township south of Mildura. Its descriptive name comes from the colour of the banks of the Murray River.

Red Hill *ACT* is a name that has been associated with the hill since the time of the early settlers.

Redcliffe *Q* is a locality in the Brisbane metropolitan area and was so named because of the red appearance of the cliffs there. It was the site of the first convict settlement in Moreton Bay, but trouble with the local Aboriginals and the need for more fresh water persuaded the authorities to move the settlement up the river to the present site of Brisbane.

Redfern *NSW*, a Sydney suburb, was named after Dr William Redfern, the ship's surgeon who was transported as a convict to Sydney for supporting sailors on *HMS Standard* who were protesting against the food supplied to them. On arrival in Sydney, he was sent to Norfolk Island where his experience and personality gained him the post of assistant surgeon and later, in 1802, the appointment as surgeon. In the same year Governor King granted him an absolute pardon and in 1808 Redfern was appointed assistant surgeon at Sydney.

Redhead *NSW* takes its name from the adjacent headland which projects into the sea.

Redland Bay *Q* was known by the Aboriginals as Talwurrupin meaning 'wild cotton tree'.

Redwood Park *SA* There is no record of the origin of this name but it is probably descriptive because of the large red gums that grew in the region prior to subdivision.

Regent *V* is a suburb of Preston in the Melbourne metropolitan area. It is named after the Prince Regent.

Regents Park *NSW* is a suburb of Bankstown, a city located in the western part of the Sydney metropolitan area. Regents Park received its name because in 1879 Messrs Jackson and Peck owned a property named Regents Park. Presumably it was called after the London suburb. The name was adopted for the terminal station when the branch line from Lidcombe was completed to this point.

Regentville *NSW*, a locality within the boundaries of the City of Penrith, is named after the homestead of Sir John Jamison whose large property was located on a grant dating back to 1805.

Reid *ACT* honours Sir George Reid (1845–1918), a former premier of New South Wales, who was one of founders of the Constitution, and Prime Minister from 1904 to 1905.

Reids Mistake *NSW*, is the entrance to Lake Macquarie. It got its name because like Newcastle, Lake Macquarie was discovered by accident. In 1800, Captain William Reid was bound for Port Hunter to pick up a cargo from the foreshores. He entered Lake Macquarie by mistake and picked up cargo from the foreshores there instead. The entrance to the lake has been known ever since as Reid's Mistake.

Renison Bell *T*, on the west coast of Tasmania, is Australia's largest tin mine. The name honours George Renison Bell, a prospector and mining engineer who, in 1890, was one of the first men to search for minerals in the area.

Renmark *SA* had its origin on 14 February 1887. This was the date of the agreement signed between the South Australian Government and the Canadian brothers, George and William Chaffey, in which the Canadians undertook to found an irrigation colony in South Australia.

The Chaffey Brothers selected the abandoned station of 'Bookmark' to establish their new colony. The Chaffeys named it Renmark, a native word meaning 'red mud'. In 1892 the Chaffey Brothers went into liquidation and were succeeded by the Renmark Irrigation Trust which took over the running of the project.

Research *V* is a locality in the Shire of Eltham on the outskirts of the Melbourne metropolitan area. Research refers to a local innkeeper, who, after the original find of gold in 1856, made a second search for a gold lead. His research led to a rush to the Research Gully field in 1861. It was formerly known as Swipers Gully, Wallaby Town and part of Caledonian goldfields. It was named Research in 1891.

Reservoir *V* is a suburb of Preston in the Melbourne metropolitan area. It is the site of a water storage scheme connected with Yan Yean and the Maroondah Dam, built in 1864.

Revesby *NSW* is a suburb of Bankstown, a city located in the western part of the Sydney metropolitan area. It was named after Sir Joseph Banks's estate in England. He was the Squire of Revesby.

Reynella *SA* is named after John Reynell, who made the first wine in South Australia. Reynell took up land in the area in 1838. He established the first vineyard in 1840 with cuttings from Sir William Macarthur of New South Wales. He laid out the township in 1854. Thomas Hardy, the founder of Hardy's wines, worked as a boy for John Reynell.

Rhodes *NSW* is a locality in the Sydney municipality of Concord. It gets its name from a house of that name built by a Thomas Walker who was an early settler at Uhrs Point. Walker chose the name because Rhodes, near Leeds, was his grandmother's home. He was a different Thomas Walker from the man who built Yaralla forty years later.

Richardson *ACT* is named after the novelist, Ethel Florence Lindesay Richardson (1870–1946), who is best known for her book *The Fortunes of Richard Mahoney* published in 1930 under the pen name of Henry Handel Richardson.

Richmond *NSW* was named by Macquarie after Richmond on the River Thames in England. Macquarie felt both places were beautifully situated.

Richmond *T* was named by Governor Sorell in 1824 when he purchased David Long's estate, Richmond Downs. In 1803 Lieutenant John Bowen explored the district and discovered coal, which led to its being called Coal River. A settlement was formed in 1815, and renamed in 1824.

Richmond *V* is one of Melbourne's oldest suburbs. The town developed rapidly after gold was discovered in 1851 because land was cheaper than in central Melbourne and the building regulations were less strict. Local government began in the area in 1855. Richmond was proclaimed a town in 1872 and became a city in 1882. It is named after the London suburb of Richmond Hill.

Ridley *SA* is located on the east and west side of Wasleys railway station at Mudla Wirra which is west of the Barossa Valley. It was laid out in 1873. Lots were disposed of privately on application to Nathaniel Oldham in Adelaide.

Ridleyton *SA* is a suburb of Hindmarsh in the inner metropolitan area of Adelaide. It is named after one of the most famous early settlers in Hindmarsh, John Ridley, who arrived in 1840 and established a flour mill. He later became known for his famous reaping machine which he invented in 1843. His machine eventually revolutionised the young Australian wheat industry by speeding up reaping and reducing harvesting costs. John Ridley purchased the land from Osmond Gilles in 1842 and lodged a plan of subdivision for Ridleyton in 1873.

Ringwood *V* is a city about thirty kilometres east of the centre of Melbourne. Ringwood was named by the early surveyors who were born near the town of Ringwood on the edge of the New Forest in Hampshire, England. Ringwood was cut off from Lillydale Shire and created a borough in 1924. It was proclaimed a city in 1960.

Ripponlea *V* is a locality of the City of St Kilda in the Melbourne metropolitan area. It takes its name from 'Ripponlea', a magnificent mansion built in Elsternwick in 1868 for Frederick Thomas Sargood, an importer and politician.

Risdon *T* was named by Captain John Hayes in February 1792, probably after the second officer on his ship, the *Duke of Clarence*.

Riverina *NSW* Several of the largest rivers flow through this extensive area in southern New South Wales. They include the Murray, Murrumbidgee, Lachlan and Edward.

Riverstone *NSW*, a locality in the City of Blacktown on the western outskirts of Sydney, was named after Lieutenant-Colonel Maurice Charles O'Connell's grant of land which he received from Governor Macquarie on the event of his marriage to Mrs Mary Putland. He named the property 'Riverston Farm' after his birthplace in Ireland. With the coming of the railway line an 'e' was added to Riverston.

Rivervale *WA* was known prior to 1884 as Brandon Hill after an early resident in the area. The name Rivervale was first used for the railway station on the Perth to Armadale line and was later adopted as the name for the postal district. The name is descriptive of its proximity to the Swan River.

Riverview *NSW* is the popular name for St Ignatius College, one of Sydney's best known Roman Catholic Schools. It is named after 'Riverview Cottage' owned by Manuel Francis Josephson who lived there from about 1866 to 1876. The name, Riverview, is now applied to St Ignatius College which was founded in 1880 by Father Joseph Dalton (1817–1905). The first college regatta was held on the Lane Cove River in 1885 and has been held annually ever since.

Riverview, Mount *see* **Mount Riverview**

Riverwood *NSW* is a suburb of the Sydney municipalities of Canterbury and Hurstville. When the railway to East Hills was opened in 1931, the station was named Herne Bay after Herne Bay, a small arm of Salt Pan Creek. Because 'the present name has been made infamous by an "emergency" housing settlement bearing the same title', a number of investigations and ballots took place. Riverwood was selected as being the most suitable popular name in 1957.

Rivett *ACT*, is called after Sir David Rivett (1885–1961), an Australian scientist who was chairman of the council for the CSIRO from 1946 to 1949.

Rivoli Bay *SA*. The bay was discovered by Nicolas Baudin in 1802. It was named after the Duke of Rivoli, whom Napolean had nicknamed 'the favoured child of Victory'. The Aboriginal name of the northern portion of the bay was Wirmal-Ngrang; the southern, Wilichum.

Robe *SA* was named after the Governor, Lieutenant-Colonel F.H. Robe, who arrived at Guichen Bay in the Government schooner *Lapwing* in January 1846, and selected the site of the town.

Robertsons Point *NSW,* a locality in North Sydney at the tip of Cremorne, bears the name of the original owner of the whole of Cremorne, James Robertson, who was granted land there in 1823 in recognition of his work in repairing scientific instruments at the Parramatta Observatory.

Robinvale *V,* a township eighty kilometres south-east of Mildura by the Murray River, was named after Robin Cuttle, a local farmer's son who was killed during World War I.

Rochdale *Q* is named after the Roche family, who took up land in the area in the 1870s to grow grapes and oranges.

Rochester *V,* a rural township thirty kilometres south of Echuca by the Campaspe River, was originally called Rowechester after Dr Rowe, the first settler in the district. It was later changed to Rochester by the Lands Department officials, who supposed it to be a misspelling of Rochester, England.

Rockdale *NSW* is a municipality, formerly known as West

Botany, in the metropolitan area of Sydney. Rockdale was the name originally given to the area at the intersection of Bay Street and Rocky Point Road in 1878 by Mrs Mary Anne Geeves, who superintended the post office attached to the general store there, but it was only really accepted after the building of the railway station bearing that name in 1884. It is first mentioned in the records of the West Botany Council in March 1886, when a letter was received from the Rockdale Progress Committee asking for attention to a long list of streets in the area. In November 1887 the Council asked the Government to change the name of the municipality to Rockdale and on 17 May 1888, the change of name became law.

Rockhampton *Q* is a city about 480 kilometres north of Brisbane. In 1856 William Henry Wiseman, a land commissioner, was sent by the Government to find a suitable site on the Fitzroy River for a township. He named it Rockhampton meaning 'town near the rocks' in the river, because of the rocky bar.

Rockingham *WA* received its name from the ship, *Rockingham,* which ran aground on the beach which fronts present day Rockingham. The town of Rockingham had its beginnings in 1846 when the Surveyor-General, J.S. Roe, surveyed the Safety Bay area with a view to establishing a timber port there. Roe advocated Mangles Bay as a better site and, following this, Surveyor Hillman surveyed a town-site in March 1847. The town-site of Rockingham was gazetted on 11 June 1847.

Rocklea *Q* is a locality in the Brisbane metropolitan area which got its name because of the rocky waterholes nearby.

Rocks, The *see* **The Rocks**

Rocky Point *NSW,* a suburb of the Municipality of Kogarah, is a rocky outcrop where the Georges River flows into Botany Bay. One of the earliest homes in the district was built there by Robert Cooper after his marriage to Catherine Rutter in 1830. It was then called Charlotte Point and the headland was part of forty hectares granted to Catherine and named in honour of her mother. Rocky Point was the popular descriptive name, and when convicts cleared a track from St Peters, town residents could take the ride to Georges River along the embryonic Rocky Point Road. Eventually a Rocky Point Road Trust was established and tolls were charged, but Rocky Point Road today has been foreshortened. Rocky Point became the name for the district south of Cooks River, and along the road to Georges River. Later there was an inn called Prendergast's Rocky Point Hotel which stood on about seven hectares of land near the junction of Rocky Point and 'Kuggerah' Roads, near Muddy Creek (now part of Princes Highway).

Rodd Island *NSW,* in Iron Cove in Sydney Harbour, is named after the Rodd family of Drummoyne. The island became part of Sydney Harbour National Park in 1981.

Rodd Point *NSW* is a suburb of the Sydney municipality of Drummoyne. It is named after the Rodd family of Drummoyne. The family was buried in a mausoleum carved out of a huge rock at Vault Point. A cross cut from a single piece of sandstone by convicts was removed to Rookwood Cemetery in 1903 but was returned to its original site in 1975 by the Five Dock Rotary Club.

Roebourne *WA* was named after Western Australia's first Surveyor-General, John Septimus Roe when the town was proclaimed in 1866. The town-site was gazetted on 17 August 1866.

Roebuck Bay *WA* was named by Phillip Parker King who surveyed the area in 1821. King called it after the *Roebuck,* the ship in which Dampier had visited the coast in 1699.

Rogans Hill *NSW* is a locality in the Baulkham Hills Shire on the north-western outskirts of Sydney. It was formerly known as Castle Keep and Bayly's Corner. The name Rogans Hill came into being when the Parramatta-Rogans Hill Railway was opened. It was named after a man called Rogan who was an early pioneer in the immediate area.

Rokeby *T* recalls Rokeby Court, now a private house. It was the original watchhouse but has also been used as a police station, courthouse and gaol. It dates from around 1840. 'Rokeby House', a residence built by George Stokell, a merchant and farmer, also dates from about 1840.

Roland, Mount *see* **Mount Roland**

Roleystone *WA* is an area in Armadale, a town on the outskirts of the Perth metropolitan area. The name was once thought to come from Roleystone Manor, an estate near the birthplace of Thomas Buckingham who bought land in the district in 1887. But research has shown that the district was already called Roleystone before Buckingham arrived. It is more likely that the name comes from an unusual rock formation which has a stone perched above it looking as if it will roll down the hill at any moment. A farm named 'Roleystone' was purchased by Thomas Buckingham in 1858, east of Kelmscott, thirty kilometres from Perth. 'The Rolling Stone' is mentioned as a place in the ranges 'near

Buckingham's' in a report by Mounted Constable Joseph H. Armstrong dated 20 November 1865. On 10 November 1865, P.C. William Regan of the 36 mile Police Station on the Albany Road reported tracks of the notorious Moondyne Joe having been found 'towards rowly stone on the Canning'.

Roma *Q* originated at Mount Abundance station. S.S. Bassett established his vineyard at Romaville in 1863 and produced the first wine to be sold in Queensland. The town was gazetted in 1867 and named after Diamantina Roma, wife of Sir George Bowen, the first governor of Queensland.

Romsey *V*, a township sixty kilometres north of Melbourne, is named after Romsey in Hampshire, meaning 'Rum's Island'.

Rookwood *NSW* is a suburb of Auburn and Strathfield municipalities. In the early days, it was known as Haslam's Creek, there being a creek which ran on the boundary of the cemetery and emptied itself in to Duck River. The area was once wild bush used for grazing purposes, but was established in 1868 as a burial ground. The suburb is still best known for its cemetery. Rookwood comes from Ainsworth's novel of that name according to the Railways Department. However, a press correspondent in 1898 said he had proposed the name of Rokewood, and that Mr Copeland Bennett said 'How would Rookwood do?'

Rooty Hill *NSW*, a locality in the City of Blacktown on the western outskirts of Sydney, was the centre of government administration for the area, and was known as Rooty Hill since at least 1810. It may be that 'rooty' refers to roots found there. But it has been suggested that it comes from *rooty* or *roti*, the Hindu term for bread which was used by Indian army men to mean tucker. Former Indian army men may have applied the name because wheat was grown or food distributed there.

Roper River *NT* was named by Leichhardt on 19 October 1845 for the first member of his party to sight the river, twenty-four-year-old John Roper. The expedition was on its way from Darling Downs to Port Essington. Roper was wounded by Aboriginals during the journey. Roper's Lake was also named after him.

Rose Hill *NSW*, a suburb of the City of Parramatta, was the name given to part of the subdivision of Elizabeth Farm and is not connected with Parramatta's original name bestowed by Governor Phillip. Sir George Rose was one of the secretaries of the British Treasury and later treasurer to the Navy.

Rosebery *NSW,* a Sydney suburb, was named in honour of Archibald Phillip Primrose, the 5th Earl of Rosebery, who was Prime Minister of England in 1894 and 1895. He spent two months in Australia with his wife in 1883 and 1884.

Rosebery *T* is named after the Earl of Rosebery.

Rosebud *V,* a bayside resort on the Mornington Peninsula, takes its name from a schooner which was wrecked in the area in the early 1850s.

Roselands *NSW,* is a suburb of the Sydney municipality of Canterbury. It was a later name for the land around 'Belmore House', bought by Stanley Parry in 1943. After World War II, the land became Roselands Golf Course, and the house became the clubrooms.

Roseneath *T* was named by Governor Macquarie during his visit to the district in 1821. 'Roseneath House', which was built in 1843, was destroyed in the 1967 bush fires, and the name is now perpetuated in the nearby Roseneath primary school.

Roseville *NSW,* a suburb of Ku-ring-gai Municipality, was the name of a stone cottage on the site of the station in an orangery. The cottage was demolished when the railway came.

Rosny *T* was a family name of W.A. Bethune, the holder of the original grant, and it was named after his ancestor, Duc de Maximilien de Bethune Sully, of Rosny near Nantes.

Ross *T* was named by Governor Macquarie in 1821 in honour of the seat of his friend, H.M. Buchanan of Loch Lomond in Scotland.

Rosslyn Park *SA,* according to South Australian nomenclature expert, the late Rodney Cockburn, is of Scottish origin and commonly spelt Roslin. There is an Earl of Rosslyn. 'Roslin Castle', which was the original residence of the noble family to whom it still belonged in 1930, is situated near a place where the River North Esk runs over a very rugged and sloping channel, emphatically called to this day, 'the Lynn'.

Rossmore *NSW,* a suburb of Liverpool, is one of the oldest agricultural districts in the state. It was originally called Cabramatta but this name was later transferred to an area which is now in the Fairfield city boundaries.

Rostrevor *SA* is a suburb of the corporate city of Campbelltown in the Adelaide metropolitan area. Rostrevor takes its name from a mansion called 'Rostrevor Hall' built by Ross Reid, who arrived in South Australia with his father in 1839 at the age of six. He married Lucy, the daughter of John Reynell of Reynella. The mansion was purchased by the Roman Catholic Church in 1923 and is now used as a residence by the Christian Brothers. The first building sites were auctioned in 1913. Rosstrevor (with two esses) is a popular seaside summer resort in County Down, Ireland.

Rothbury *NSW* is named after a village near Morpeth, England.

Rottnest Island *WA* It has generally been thought that the island was named Rottnest by Willem de Vlamingh between 29 December 1696 and 2 January 1697, while he was visiting there. However the only reference to the name of the island in Vlamingh's log is '. . .saqen het mist eilandt' (. . .saw the island of mist). This was later mistranslated by a French journalist as 'Island of Girls' (mist/miss) believing it to be an island of an entirely female population. The present name was evidently coined by a Dutch maritime clerk who interpreted Vlamingh's reference to 'Wooderats' (quokkas) as meaning 'rats'.

Rouse Hill *NSW*, a locality in the City of Blacktown on the western outskirts of Sydney, commemorates Richard Rouse (1774–1852) who arrived in Sydney in 1801 and was appointed Director of Government Building in Parramatta. He named his grant Rouse Hill. The district was known as Vinegar Hill until the 1850s when the name was changed.

Rowville *V* is a suburb of the City of Knox on the outskirts of the Melbourne metropolitan area. Rowville was named about 1903 on the suggestion of a local blacksmith, as a compliment to the Row family whose property, 'Stamford Park', had been important in the district since the 1860s.

Roxby Downs *SA* is named after Lieutenant H. Roxby, RN who from 1863 made surveys along the South Australian coast. It is a pastoral lease area, and the site of a large uranium deposit at Olympic Dam.

Royal National Park *NSW* was founded by Sir John Robertson, Premier of New South Wales, as The National Park, when land was set aside as a public reserve in 1879. The term 'Royal' was adopted during the visit of Queen Elizabeth II to Sydney in 1954.

Royal Park *SA* is a suburb of Woodville, a city in the Adelaide metropolitan area. According to Rodney Cockburn, Royal Park was a fanciful name bestowed by T.J. Matters. For many years it was more accurately known as Piggery Park because of the farms and slaughterhouses there. Even in the 1950s it was still mostly swamp, scrub and mosquitoes — as many migrants found when they bought the cheap land and built temporary homes while they saved to build permanent houses.

Rozelle *NSW* obtained its name from the innermost portion of White Bay, known in the mid-1800s as Rozella Bay, because of the abundance of parrots.

Ruined Castle Valley *Q* in the Carnarvon Ranges was named in 1844 by Ludwig Leichhardt who was reminded of ruined castles in Germany when he saw the rock formations in the valley.

Rum Jungle *NT*, about 100 kilometres south of Darwin, was the site of a uranium discovery in 1949. There is no jungle. The area is full of rocky outcrops. The name was applied about an early hotel (1872) there which ran only to rum in serving its customers.

Rumney, Mount *see* **Mount Rumney**

Ruse *NSW* honours the former convict, James Ruse, who grew one of the earliest wheat crops in Australia at Parramatta in 1790. He spent his last years near Campbelltown and was buried there. The first grist mill built in the Campbelltown district was constructed by Ruse and is still there.

Rushcutters Bay *NSW* is located on Sydney Harbour in the Municipality of Woollahra. Rushcutters Bay dates from the early days of the settlement when parties of rushcutters were sent there to collect reeds for thatching huts. On 30 May 1788 two convicts, William Okey and Samuel Davis, were killed by Aboriginals while cutting rushes.

Rushworth *V* is a rural locality twenty-nine kilometres west of Murchison near the Waranga Basin. When gold was discovered there in 1853, it was originally known as the Wet Diggings. It was later changed to Rushworth, reputedly from the remark of a goldfields warden, who said it was 'a gold rush worth while'.

Russell *ACT* was the name given by Surveyor Scrivener about 1910 to a nearby trigonometrical station. The name was later adopted as an early name for the locality.

Russell Lea *NSW* is a suburb of the Sydney municipality of Drummoyne. It takes its name from an estate owned by Russell Barton. The estate was 60 acres (24 hectares) and was called 'Russell Lea' in 1900. It was first subdivided in 1913.

Rutherglen *V* is a township surrounded by vineyards forty kilometres north of Wangaratta. Its name comes from the birthplace in Scotland of John Wallace, a notable figure in the days of the goldfields.

Rydalmere *NSW* originated from Rydal in England and was part of the Old Vineyard Grant subdivision of 1886.

Ryde *NSW* is a municipality in the Sydney metropolitan area and was named after the town on the Isle of Wight, across the water from Portsmouth. Reverend George Turner, who was appointed to St Anne's Church in 1830, came from Wiltshire, but his wife Anne was a native of the Isle of Wight. Another Isle of Wight family, the Popes, moved to the district shortly after the Turners. George Pope combined shoe-making with a general merchandise shop called the Ryde store.

Sadleir *NSW,* a suburb of Liverpool, is named after Richard Sadleir, Liverpool's first mayor.

Safety Bay *WA* was noted by Surveyor-General John Septimus Roe in 1837 as a 'safe well protected boat anchorage'. He gave it the appropriate name of Safety Bay.

Saint Arnaud *V,* a township seventy-five kilometres north of Ararat, was first known as New Bendigo. It became a goldfield during the Crimean War, and was named after the French commander of the Crimea, Jacques Leroy de Saint Arnaud. It was proclaimed a borough in 1863 and a town in 1950.

St Clair, Lake *see* **Lake St Clair.**

Saint George *Q* was named by the explorer, Sir Thomas Mitchell, after he had crossed the Balonne River at a nearby ford on St George's Day, 23 April 1846.

St Ives *NSW* It has been suggested that this is named after the St Ives in Cornwall. But a more likely explanation, and one that is corroborated by official post office records, is that it was a compliment to Isaac Ellis Ives, the Member of Parliament for St Leonards from 1885–1889, and afterwards Mayor of North Sydney. In June 1885, a petition for a post office was presented by Phillip Fletcher Richardson and Mr Ives. Mr Richardson suggested Rosedale, the name of the original grant made to Daniel Mathew. But, as there was already a Rosedale post office in New South Wales, Mr Richardson, on behalf of the residents, changed it to St Ives.

Approval was granted on 10 October 1885, and the first postmaster, Sydney Smith, took up duty a month later.

St Johns Park *NSW* is a suburb of Fairfield on the south-western outskirts of Sydney. It is part of the old parish of St John. In former times, one seventh of each parish had to be set aside as a corporate school and church land. *Wells Maps and Gazetteer* of 1848 shows St John's Farm where this district is now located.

St Kilda *V* is a city on the shores of Hobsons Bay about five kilometres south of the centre of Melbourne. St Kilda is probably named after the ship *Lady St Kilda* which was visiting Melbourne in 1841 when the first land sales took place. St Kilda became a borough in 1863. In 1890 it was proclaimed a city.

St Leonards *NSW*, a locality in North Sydney, was named after the place in England of the same name by the explorer, Sir Thomas Mitchell, when he explored the area in 1828. For many years the whole North Sydney area was known by that name but it is now reserved for the area around the railway station and St Leonards Park in North Sydney, several kilometres from St Leonards station.

St Leonards *T* was formerly Patersons Plains. St Leonards was proclaimed as a town on 3 July, 1866. K.R. von Stieglitz in his book *A Short History of St Leonards* says that it is supposed to have been named after Lord St Leonards (born Edward Sugden 1781–1875). There is a resort town of St Leonards near Hastings in Sussex.

St Marys *NSW*, a locality within the boundaries of the City of Penrith, is named after the parish church of St Mary Magdalene, consecrated in 1840. It was earlier known as South Creek.

Saint Mary's Peak *SA* The appropriateness of the name of the mountain, which was sighted by E.J. Eyre in 1840 and named by B.H. Babbage in 1856, has been disputed. C.H. Harris says, 'the appropriateness of the apellation is obvious to anyone acquainted with its appearance and surroundings, for the white mantle of snow by which it is more frequently invested than any other hill in South Australia conveys an irresistible impression of saintliness' — a reference to Mary, the mother of our Lord. However, snow is rarely seen on the mountain. In fact, Mr Hunt, lessee of Wilpena Pastoral Run, said in 1985 that he had only seen snow once in the last sixty years or so. He recalled that on this occasion, in the early 1950s, snow stretched from the valley floor to halfway up the peak on one side. There was no snow on the actual crest.

Hans Mincham in *The Story of the Flinders Range* draws attention to the fact that Babbage named St Mary's Pool in the MacDonnell Creek, and lived at St Mary's, Adelaide, both of which may have some significance in relation to his naming of the peak.

St Peters *NSW*, a suburb of Marrickville, was named after the old church of the same name which was built in 1838 on the Cooks River Road (now Princes Highway) and which still stands, now without its upper steeple. Many gravestones of the earliest settlers in the district are to be found in the adjoining graveyard.

St Peters *SA* is an inner suburb and a corporate city of Adelaide. Its name is derived from the Collegiate School of St Peter, one of Adelaide's oldest and best-known private schools. It opened in 1847 and moved to the area a few years later. The school's foundation stone was laid by Bishop Short on 24 May 1849. Bishop Short was educated at St Peter's College, Westminster in England. The corporate town was established in 1883.

St Vincent Gulf *see* **Gulf St Vincent.**

Salamander Bay *NSW* was named after the *Salamander,* the first boat to enter the port. The bay was once proposed as the site for a naval base and for heavy industry.

Sale *V* is a city in East Gippsland near the junction of the Latrobe and Thomson Rivers. Sale was first discovered by Angus McMillan in 1841. The district was first settled in 1844 by Archibald McIntosh. Governor Fitzroy approved the plan for the village of Sale in 1850. It is named after Sir Robert Henry Sale, a distinguished British army officer, who was killed in the Afghan border wars in 1845. Sale was proclaimed a town in 1924, and a city in 1950.

Salisbury *SA* is a city on the northern outskirts of the Adelaide metropolitan area. Salisbury was named by an early land owner John Harvey, who developed the village and began selling town blocks in 1848. He named the village after the country town of his wife's English home in Wiltshire.

Salt Ash *NSW* was an early Port Stephens settlement site, supposedly named after its counterpart British place name. It was once a busy road–water transport connection point and a timber mill used to be situated there on Tilligerry Creek.

Samford *Q* was the name of a station established by John Delaney Bengin.

Sandgate *NSW* was a sandy hill and a tract of waste land when the Newcastle Council acquired it and named it Sandgate after an English town.

Sandringham *NSW,* a suburb of the Municipality of Rockdale, was the name given to the area surrounding the famous Prince of Wales Hotel.

Sandringham *V* is a city of residential suburbs built around Port Phillip Bay about twenty kilometres from the centre of Melbourne. The town of Sandringham was originally known as Gipsy Village because a party of gipsies lived there. In 1888 it was renamed Sandringham after the royal estate in Norfolk, England, by C.H. James, a local landowner. The district was part of Moorabbin until the Shire of Sandringham was declared in 1917. Sandringham became a city in 1923.

Sandy Bay *T* was named by the Revd Robert (Bobby) Knopwood, who is supposed to have applied this name descriptively to the water feature. The name of the suburb followed.

Sandy Hollow *NSW* is probably a descriptive name.

Sans Souci *NSW,* a suburb of the Sydney municipalities of Kogarah and Rockdale, was the name of a mansion built by Sir Thomas Holt, MLA, on land acquired at Rocky Point in the mid-nineteenth century. However, his German-born wife did not like the wilds of Rocky Point so the mansion, named in honour of the famous Potsdam palace built by Frederick the Great in 1745, was sold about 1865 to a restaurateur named William E. Rust and became Sans Souci Hotel. The beautiful situation, the mild climate, the name (which means 'carefree') established the area as a notable seaside resort which persisted even after the Sans Souci Hotel was demolished.

Sarina *Q,* was a name bestowed by a Greek surveyor.

Sassafras *V* is a locality in the Shire of Sherbrooke on the outskirts of the Melbourne metropolitan area. Sassafras is named after the sassafras trees growing there. Sassafras Gully was the name chosen for the post office established by a settler named Arthur Goode in 1894.

Savage River *T* is a fairly long, northern tributary of the Pieman River on the West Coast. The name was given by Surveyor Gordon Burgess in honour of one of his companions, Job Savage, in the exploration of the country south and west of Mount Cleveland in 1864. There is also a town named Savage River situated about twenty-seven kilometres west south-west of Waratah. This was established in the mid 1960s, and services the mine of this name from which iron ore is piped to Port Latta on the North-West Coast, where it is pelletised for shipment.

Sawpit Creek *NSW* is known to Perisher Valley skiers as the site of a car park. It gets its name because timber getters used to haul logs there and place them over large pits. A saw was then operated with one man standing on the log sawing down, and one standing in the pit, sawing up. Some of the old bullock tracks used by the timber getters hauling their logs to Sawpit Creek are still visible along the Thredbo River Valley.

Scarborough *NSW,* a suburb of the Municipality of Rockdale, is named after the English coastal town and is the name of the original subdivision.

Scarborough *WA* in the city of Stirling in Perth metropolitan area shares its name with Scarborough in Yorkshire. Patrick Callaghan bought up land in this area in October 1892. It was soon after that he subdivided and a few blocks were sold. It is not known if he advertised the land as 'Scarborough' although it is possible to attribute the origin of the name to him. He named one of the streets in his subdivision Scarborough Road.

Schanck, Cape *see* **Cape Schanck.**

Schofields *NSW,* a locality in the City of Blacktown on the western outskirts of Sydney, was named after a local family. The Schofields bought a small area of land in the area after they sold their grant at Marsden Park. Schofield's Siding was built in 1870 on this land.

Scone *NSW* was named after the village in Perthshire, where the ancient kings of Scotland were crowned. In 1831, an early settler, Hugh Cameron, presented a petition asking that the name, Strathearn, should be used, mentioning that the valley of Strathearn was close to the 'Ancient Place and Abbey of Scoone'. The village was gazetted as Invermein in 1837, but this was later changed, under instructions from the Colonial Secretary, to Scone.

Scoresby *V* is a suburb of the City of Knox on the outskirts of the Melbourne metropolitan area. Scoresby was named in honour of the Revd W.S. Scoresby, a doctor of Divinity better known as a distinguished marine surveyor, who visited Australia in 1856. His death in England in 1857 coincided with the survey of the district which was named after him.

Scotland Island *NSW,* the only island in Pittwater, was named in memory of the native land of Andrew Thompson, a former convict who was granted the land by Governor Macquarie in 1810. Thompson died a year after receiving the grant.

Scottsdale *T* is named after James Scott, a government surveyor, who cut his way through the dense scrub of the North-East in 1852.

Scullin *ACT* honours James Henry Scullin (1876–1953), who was Prime Minister from 1929 to 1931.

Seacliff *SA* is a descriptive name.

Seacombe Heights and Seacombe Gardens *SA* are suburbs of the City of Marion in the southern part of the Adelaide metropolitan area. The land was subdivided privately and named after a place in Cheshire. Seacombe is now officially included in the suburb of Seaview Downs.

Seaforth *NSW,* a suburb of Manly, has Scottish associations in Loch Seaforth and Seaforth Island. The area now known as Seaforth was divided into numerous allotments by its owner, Henry F. Halloran. The first auction sales were held on 10 and 11 November 1906. Houses built there subsequently have some of the best harbour views in Sydney.

Seaton *SA* is a suburb within the corporate city of Woodville, in the Adelaide metropolitan area. Seaton was auctioned, without success, in 1883. The bulk of the section was bought much later, in 1906, by the Adelaide Golf Club for golf links at eight pounds an acre. Seaton is a common name in Scotland and England, and may have been named by W.G.P. Joyner or Gifford Tate, a former secretary of the golf club, after his birthplace. Seaton is today a championship golf course and is still the headquarters of the Royal Adelaide Golf Club. Seaton Park was formerly called Grangeville.

Sedgwick *V* was a township fourteen kilometres south of Bendigo. The rural district was known until 1903 as Upper Emu Creek, after which it was changed to Sedgwick after Adam Sedgwick, the British geologist.

Sefton *NSW* is a suburb of Bankstown, a city located in the western part of Sydney metropolitan area. It got its name because in 1839 James Woods called his grant of land by that name.

Semaphore Park *SA* is a suburb within the corporate city of Woodville, in the Adelaide metropolitan area. It was intended as a fashionable seaside town. Semaphore Park was originally New Liverpool, then Mellor Park (after the landowner Joseph Mellor). Semaphore was so named because it was the site of a signal station and landing place.

Seppeltsfield *SA* At the site of the largest family-owned winery in the world, Joseph Seppelt planted his first vines near Nuriootpa in 1851. The railway station, Seppelts, at Seppeltsfield was changed to Dorrien after General Smith-Dorrien, a British leader in World War I, when many place names of German origin were changed.

Serviceton *V*, a small township near the Victorian and South Australian border, is named after James Service, a former premier of Victoria.

Seven Hills *NSW* is a locality in the Baulkham Hills Shire and the City of Blacktown on the north-western outskirts of Sydney. Seven Hills owes its name to Matthew Pearce, one of the first free settlers to come to New South Wales. From his house, seven hills were visible — hence the name. Pearce arrived with his wife aboard the *Surprise* in 1790. He died in 1831 at the age of sixty-nine. The Pearce family has occupied Kings Langley for over 130 years. The National Trust has listed the Pearce graves in Seven Hills Road.

Seven Shillings Beach *NSW* is located on Sydney Harbour in the Municipality of Woollahra. Seven Shillings Beach got its name because when Mr W. Busby bought 'Redleaf' in 1871 an Aboriginal named Gurrah had fishing rights to the beach and lived just outside Redleaf's fence. The Busbys had continual trouble with members of the tribe so Mrs Busby tried to buy the fishing rights and offered them blankets and flour and clothes. At last Gurrah said he would sell for seven shillings and eventually Mrs Busby gave Gurrah the money and the tribe moved.

Seventeen-Seventy *Q* This unusual name comes from the fact that Captain Cook landed in the area in that year. The name was given official approval in 1953.

Seymour *V*, a township by the Goulburn River ninety-eight kilometres north of Melbourne, got its name when the town-site was surveyed in 1843. It was named after Lord Seymour, son of the 11th Duke of Somerset, and a minister in Lord Melbourne's cabinet.

Shalvey *NSW*, a locality in the City of Blacktown on the western outskirts of Sydney, is named after the main road in the area. This road was originally a crown subdivisional road.

Shark Island *NSW* is a reminder of the sharks that make it dangerous to swim in Sydney Harbour. The Aboriginals called the island Boambilli.

Shay Gap *WA* is named after R. Shea, who with S. Miller was a pearler from Ninety Mile Beach. As was the custom, they recruited Aboriginal divers from the area. In January 1873, they left their camp near the De Grey River to look for certain deserters. They were speared by the Aboriginals, and their mutilated bodies were discovered near Muccanoo Pool in the De Grey River a month later. The name, Shay, was given to the gap twelve kilometres north-north-east of where the murders occurred. The nearby private townsite of Shay Gap named after this feature was approved on 25 November 1970.

Sheffield *T* was probably named by Edward Curr, the first local manager of the Van Diemens Land Company, after his home in England.

Shellharbour *NSW*, was explored by George Bass and Matthew Flinders in 1797. They gave the harbour its name because of its shell-strewn beach. It was proclaimed a municipality in 1859. The original spelling, Shell Harbour, has been amended to Shellharbour.

Shelly Beach *NSW*, a locality in Manly, was so called, not after the poet Shelley, but because until comparatively recent times it was always thickly covered with shells.

Shepparton *V*, a city on the Goulburn River 198 kilometres north of Melbourne, was known as Macguire's Punt. The name was inscribed on nearly all documents, particularly government records. The name Shepparton first came into use in 1855, apparently with no relation to the town in England called Shepperton. The pioneers in early correspondence called the town Sheppardtown or Sheppardton after Sherbourne Sheppard, a squatter at Tallygaroopna station. Shepparton was proclaimed a township on 28 September 1860, when the name, Macguire's Punt, was dropped.

Sherbrooke *V* is a shire in the Dandenong Ranges centred on Upwey. It is known for its beautiful forests and lyrebirds. It is also known as the home of *Puffing Billy,* a restored steam engine of the early 1900s which takes passengers from Belgrave to Emerald. Sherbrooke derives its name from the Canadian birthplace of R.W. Graham who succeeded in having a post office established there in 1894 and remained as postmaster for the next twenty-five years. Sherbrooke was part of the original Shire of Fern Tree Gully when it was formed in 1889. It was renamed in 1964 when part of Fern Tree Gully Shire was cut off to form the City of Knox.

Shoal Bay *NSW* was supposedly named by Governor Lachlan Macquarie because of the sand shoals that exist there.

Shoalhaven River *NSW* was named Shoals Haven by George Bass in 1797.

Shortland *NSW*, a suburb of Newcastle, is named after Lieutenant John Shortland of HMS *Sirius* who, on 8 September 1797, discovered the estuary of the Hunter River while searching for escaped convicts.

Simpson Desert *NT SA* This vast stretch of arid land was first entered by Charles Sturt in 1845, but was not named until 1929. It was named by Dr C.T. Madigan, for A.A. Simpson, president of the South Australian branch of the Royal Geographical Society. Simpson was largely instrumental in raising the funds for an exhaustive aerial survey which Dr Madigan conducted, the results being published in his book *Crossing the Dead Heart*.

Singleton *NSW* took its name from Benjamin Singleton who was a member of a European discovery party led by John Howe, chief constable of Windsor, in 1820. In 1822, Singleton settled on the present site of the town. The town's first jail, which he built in 1841, still stands. It now houses the Singleton Historical Museum.

Skirmish Point *Q* was named by Matthew Flinders in 1799 after an Aboriginal tried to seize Flinders' hat and threw a spear. Flinders fired and wounded the spear thrower.

Slack's Creek *Q* was named after Mr W. Slack who had a cattle run there before permanent settlers arrived. The Aboriginal name was Mungaree meaning 'place of fishes'.

Smiggin Holes *NSW* is the name of a ski resort. It is a Scottish term referring to the holes worn in the ground by stock around a salt lick.

Smithfield *NSW* is a suburb of Fairfield on the south-western outskirts of Sydney. Smithfield is mentioned in newspaper references covering the development of the area. The town had several names (Sherwood, Prospect and Donnybrook) before the name of Smithfield was finally chosen and this came about when in 1836 John Ryan Brennan purchased a tract of land upon which he built a large cattle yard and meat market. Because the major meat markets in both London and Dublin were called Smithfield, it was decided to call Brennan's venture after the British market's name.

Smithfield *SA* was named by Mr John Smith who bought land there about 1848. He built the Smithfield hotel early in the 1850s.

Smithton *T* is named after Peter Smith, who applied for five town allotments in 1855.

Smoke Creek *WA* is a watercourse about twenty-five kilometres long which flows generally from a divide twelve kilometres west of 'Lissadell' pastoral station homestead into Lake Argyle. The name is most probably descriptive.

Smoky Cape *NSW* was named by Captain Cook in 1770 because he saw smoke from Aboriginal camp fires.

Snapper Island *NSW* in Sydney Harbour is a traditional name. Anyone is entitled to guess who named it, when and why.

Snowtown *SA* was named by Governor Jervois when the town was proclaimed in 1878 after Thomas Snow, his cousin and aide-de-camp, and his brother, Sebastian Snow, who was the Governor's private secretary.

Snowy Mountains, Snowy River *NSW V* is a popular descriptive name which has supplanted the Aboriginal name, which was Muniong (or Munyang), and was bestowed on the range by the Revd W.B. Clarke, who made a geological exploration in 1851–52. The mountains had been described (though not named) as 'snowy' by Captain Mark Currie in 1823 and by Hamilton Hume in 1825. The Snowy River rises east of the Great Dividing Range in New South Wales and flows through East Gippsland to enter Bass Strait at Marlo.

Solander, Cape *see* **Cape Solander**.

Soldiers Point *NSW* was named after the Corporal's Guard once stationed there. Site of the Cromarty land grant, it was formerly called Friendship Point.

Solitary Islands *NSW* were named by Captain Cook on 15 May 1770.

Sorrento *V* is a bayside resort on Mornington Peninsula. When Charles Gavan Duffy arrived in the 1860s, he built a large house here and named it Sorrento because the landscape was reminiscent of the Italian Sorrento. It was the site of the first settlement in October 1803.

South Australia Before the official designation of the state was determined in 1836, several alternative suggestions were discussed. Dr John Dunmore Lang advocated Williamsland in honour of King William IV, and D.I. Gordon wished to call it Central State. For some time, the whole area was known as Flinders Land. Baudin's map had shown it as Terre Napoleon. The name South Australia became legally connected with the colony by Act 95 signed by William IV on 15 August 1834.

South Melbourne *V* is one of Melbourne's older areas and has many buildings of historic interest. Local Government began in the area when the Municipal District of Emerald Hill was proclaimed in 1855. Emerald Hill was proclaimed a city in 1883 and the name was changed to South Melbourne.

South Molle Island Q was named by Lieutenant Charles Jeffreys of the brig *Kangaroo* in 1815. He called it after the Lieutenant-Governor, Colonel George Molle.

South Steyne *NSW,* a locality in Manly, was renamed Manly Beach in 1976 by the Geographical Names Board. The name remains in the South and North Steyne roads along Manly Beach front and in North Steyne Post Office and North Steyne Surf Life Saving Club. The name Steyne originated from a thoroughfare in Brighton, England. Henry Gilbert Smith used the name Steyne because he was trying to model Manly on Brighton.

South Sydney *NSW,* a Sydney suburb, was originally called the Municipality of Northcote when boundaries were changed in 1968. The name was later changed to South Sydney by popular acclaim. The Municipality of South Sydney incorporated a number of areas that had their own local government at one time in their history. South Sydney was absorbed by the Sydney City Council in 1982. It was re-established as a separate city by parliament in 1988 and the new city came into existence on 1 January 1989.

Southern Cross *WA* was so named because two prospectors, T.R. Risely and M. Toomey, were directed to the hills they were seeking by being told that they should travel slightly to the east of the Southern Cross. They did so and were rewarded by finding a promising goldfield, which they named on account of their experience. When the township was laid out, the streets were named after various constellations, such as Altair, Antares, Sirius, Spica, etc.

Southport *Q* was surveyed in 1874 and named by Thomas Blacket Stephens, then Minister of Lands, after his birthplace in Lancashire. It was earlier known as Nerang Heads. The Aboriginal name was Goo-en. In 1875, Thomas Hanlon bought land in the new town and opened his hotel at Nerang Creek Heads, where Southport later developed. Southport Division Board was formed in 1883. The town of Southport was established in 1902.

Spearwood *WA* is an area of Cockburn, a city near Perth. In

1897, James Morrison subdivided his land, Location 264, naming the access road along the eastern boundary Spearwood Avenue. Presumably the name came from the bean and pea-plant supports used by market gardeners in the area. At the turn of the century the construction of a railway from Rob Jetty to the Agricultural Hall at Jandakot was approved. Land for a siding west of Spearwood Avenue was resumed in February 1905 and the siding was duly named Spearwood. The locality of Spearwood was included in postal directories in 1907 but the original Spearwood Avenue was later changed to Pearce Avenue and is now a portion of Rockingham Road.

Spectacle Island *NSW* in Sydney Harbour is a traditional name. It is said to get its name from its shape. It has been used since 1884 as a naval armament depot.

Speers Point *NSW* is named after a landholder called Speer who was also a member of parliament.

Spence *ACT* is named after William Guthrie Spence (1846–1926), a labour leader who served in the first parliament after Federation in 1901.

Spencer Gulf *SA* Early in 1802, Matthew Flinders hoped that the extensive inlet was the commencement of a channel leading northwards as far as the northern coast, thus providing a route to the centre of the continent. On 20 March 1802, when he realised that his hopes were not to be fulfilled, he named it Spencer's Gulphe in honour of George John, the 2nd Earl Spencer, who presided at the Board of Admiralty when the voyage was planned and ship put into commission. He also named Cape Spencer. Nicolas Baudin attempted to call it Golphe Bonaparte.

Spencers Creek *NSW* is a stream that flows under the main road to Charlotte's Pass near the summit of Mount Kosciusko. The stream was named by stockmen who accompanied James M. Spencer when they were taking stock on to the main range. When Spencer endeavoured to cross the swollen stream on his horse, he fell in. The stockmen catching sight of Spencer's head and his big beard floating on the water are known to have said, 'if he wants to have a swim, he can have it to himself', so they called it Spencer's Creek.

Spit, The *see* **The Spit**

Spotswood *V* is a suburb of the City of Williamstown which lies across from the City of Melbourne. Spotswood was the home of John Stewart Spotswood whose daughter in 1869 married Richard Seddon, later Prime Minister of New Zealand.

Springvale *V* is a residential and industrial city about twenty-five kilometres from Melbourne. Springvale probably comes from the fact that there were a number of springs in the vales or valleys of the district but nobody knows for certain how the name came about. In the early days the name was spelt Spring Vale and a hotel of that name was one of the earliest buildings in the district. Springvale was created a district in 1857 and proclaimed a shire in 1873. The name was changed from the Shire of Dandenong to the Shire of Springvale and Noble Park in 1955. Springvale was proclaimed a city in 1961.

Springwood *NSW*, a locality in the Blue Mountains west of Sydney, was named by Governor Macquarie in 1815 on his journey to Bathurst. He camped in this locality and a spring close by provided water for the men and horses. The Governor in his journal recorded the place 'being very pretty. I have named it Springwood'.

Springwood *Q*, was called Wire Paddock when it was bought in 1932 by Brigadier Sam Langford, a World War I veteran, who was awarded the Military Cross and Bar. He renamed the property 'Springwood' because of a spring in the middle of the land.

Sprouls Lagoon *NSW* was probably named after the one-time owner of 'Temora' pastoral property, Francis Sproul.

Standley Chasm *NT* was named after the late Mrs Ida Standley who was a teacher at Alice Springs from about 1918 to 1928.

Stanley T was laid out in 1842 and named by James Gibson after Lord Stanley, Secretary of State for the Colonies. Gibson succeeded Edward Gurr as agent of the Van Diemen's Land Company.

Stanmore *NSW*, a suburb of Marrickville and the site of Newington College, was once part of the estate of Mr John Jones who named it 'Stanmore' after his birthplace in England. It was on a corner of this estate that the first religious services took place in the district. Mr John Jones willed ten hectares of this estate to the Methodist Church. In 1880 Newington College was built on the site. Stanmore Station was opened in 1878.

Stanthorpe *Q* is a combination of *stannum* meaning 'tin', and *thorpe* meaning 'town'.

Stanwell Park *NSW* was part of a grant promised to Matthew John Gibbon in 1824. It is thought that the land was named after the village of Stanwell in Middlesex, probably in the period of Surveyor-General Mitchell.

Stawell *V*, a township in the Wimmera thirty-one kilometres north-west of Ararat, was named after Sir William Foster Stawell, a chief justice of Victoria. It was formerly known as The Reefs after gold was discovered there in 1854. It is pronounced 'stall' to rhyme with 'ball'.

Stephens, Port *see* **Port Stephens**

Stirling *ACT* honours Sir James Stirling (1791-1865), the first governor of Western Australia from 1829 to 1838.

Stirling *SA*, in the Adelaide Hills, is named after Edward Stirling, Senior, MLC, one of the founders of the Australian Constitution. One of his sons, Sir Edward Charles Stirling, became professor of Physiology at the University of Adelaide, director of the South Australian Museum, a member of parliament, and the first chairman of the District Council of Stirling. His other son, Sir Lancelot Stirling, became a lawyer and a sheep and cattle breeder and a member of the South Australian parliament. The township was laid out and christened by P.D. Prankerd who was a close friend of Edward Stirling.

Stirling *WA* is a city located to the immediate north of the city of Perth. Stirling was named in honour of Sir James Stirling (1791–1865) when the city was created in 1971. Stirling, the first governor of Western Australia, brought the first settlers to Perth in June 1829.

Stirling North *SA* is located near Port Augusta. It was surveyed by Gavin Young in 1849. He called it after Stirling in Scotland.

Stirling Range *WA* was named on 4 November 1835 by the surveyor, John Septimus Roe, after Governor Stirling, the first governor of Western Australia, who was with Roe on this expedition from Perth to Albany.

Stockinbingal *NSW* is a named after the pastoral selection of Sir James Mathieson of 1847. It is shown in the *1866 Gazetteer* as being a very small town called Stockinbingy. The word is probably Aboriginal.

Stonyfell *SA* is a suburb in the corporate city of Burnside to the east of Adelaide. It has been associated with stone quarries, olive growing and wine production. It takes its name from 'Stonyfell House', which dates back to a single-storey, bluestone cottage built in 1838 by W. Edin. In 1858 the house was bought by Henry Septimus Clark who planted the first vines there. A quarry at Stonyfell was used in the building of Adelaide Gaol.

Stradbroke Island *Q*. The nomenclature of Moreton Bay centres round Captain Henry John Rous, commander of HMS *Rainbow,* which conveyed Governor Darling there in June 1827. Stradbroke Island was named after Rous's father, Viscount Dunwich, Earl of Stradbroke. Rainbow Beach is where the *Rainbow* anchored. The channel between Stradbroke Island and Moreton Island was named for Captain Rous, and his father was also remembered in Dunwich Point.

Strahan *T*, was named after Sir George Cumine Strahan who was Governor of Tasmania from 1881 to 1886.

Strathfield *NSW* is a municipality in the Sydney metropolitan area. Strathfield replaced Redmyre as the name of the area when it was incorporated in 1885. It may have been named after 'Strathfield House' which was owned and occupied by John Hardy.

Strathpine *Q* is a locality in the Brisbane metropolitan area and gets its name from a Scottish word, *strath,* meaning 'valley'.

Streaky Bay *SA* was so named by Matthew Flinders in January 1802 because of the discoloured streaks he noticed in the wake of the *Investigator.* The earliest European contact went back to 1627 when Nuijts and Thijssen sailed along the south coast in the *Gulden Zeepaard,* and turned back at this point. The Aboriginal name was Cooeyanna.

Struck Oil *Q* A name that is redolent of the 1960s and 1970s, but in fact comes from a melodrama called *Struck Oil.* According to Bill Beatty, some prospectors who were visiting Mount Morgan to register a claim on the Dee River saw the play and decided to adopt the name. They met the leading lady, Maggie Moore. A.E. Martin adds the following information—Maggie's real name was Margaret Virginia Sullivan, who married J.C. Williamson in 1874 and helped her husband in his theatrical enterprises.

Strzelecki Ranges *V*, a mountain range in Gippsland, was named by Captain J.L. Stokes of the *Beagle* after the Polish explorer, Paul Edmund de Strzelecki, who explored Gippsland in 1840. The Australian pronunciation is Striz-leck-ee, but in Polish it is nearer to Stcheletzki.

Studley Park *V* is a suburb of the City of Kew in the

Melbourne metropolitan area. It is named after Studley in Yorkshire, England. The name came from a house built in the locality by John Hodgson.

Sturts Stony Desert *SA* 'That iron region' was crossed by Sturt in 1845. He said that the gibbers reminded him of shingle on a beach. Sturt River in South Australia is also named after him.

Subiaco *WA* is a city in the Perth metropolitan area. The first Europeans to settle in the Subiaco area were Benedictine monks who had purchased land in the area. In 1851 a small group led by Joseph Benedict Serra built a few primitive huts near the eastern side of Lake Monger. Serra, who dreamed that this would be the base from which missionary work could be carried out all over the colony, had named it New Subiaco. Subiaco derived from *sub lacum*, Latin for 'below the lakes', and was the name of the town in Italy where the Benedictine Order had been founded. Changing policy in the Vatican later made the settlement redundant and by 1864 it had been abandoned, with most of the Benedictines joining Bishop Salvado at New Norcia. When the Perth-Fremantle railway was completed in 1881, a little railway platform was built near the monastery, which had now become an orphanage and named 'Subiaco'.

Success *WA*, an area of Cockburn, a city near Perth, was named in 1973 for the ship commanded by Captain James Stirling when he visited the Swan River in 1827. It was gazetted on 23 March 1973. In 1968, the Post-master-General's department discussed the naming of the area with the Cockburn Shire Council who suggested Omeo, apparently after a ship that went ashore at Coogee around the turn of the century. Investigation showed that this name was duplicated in Victoria and eventually the name Success was proposed.

Sullivan Cove *T* was named after John Sullivan, Under-Secretary of the Colonial Office.

Summer Hill *NSW*, a suburb of Ashfield, was originally called Sunning Hill. This name was originally the name of the racecourse, but it was extended to include the pastoral hillside that formed part of the older Robert Campbell's Canterbury estate. It is not known how or why the name was changed to Summer Hill.

Sunbury *V* is a residential area in the Shire of Bulla on the outskirts of the Melbourne metropolitan area. The name came from the Sunbury Inn on Jackson's Creek, and this in turn from Sunbury-on-Thames in Middlesex and so named

266

by Samuel and William Jackson, two members of Fawkner's group of pioneers.

Sunshine *V* is an industrial city about twenty kilometres west of the centre of Melbourne. Industry came to the district when Hugh Victor McKay moved his Sunshine Harvester works from Ballarat to Braybrook in 1906. The H.V. McKay-Massey Harris Pty.Ltd. grew to become the largest agricultural manufacturer in the Southern Hemisphere. The town changed its name from Braybrook to Sunshine in 1907 and in 1951 the Shire of Braybrook became the City of Sunshine.

Surfers Paradise *Q* was known by the local Aboriginals as Kurrungel, because of the plentiful supply of a certain hardwood tree of that name that was useful for making boomerangs. The area was known as Elston when James Cavill, a visitor from Brisbane, decided to build the Surfers Paradise Hotel there in 1923. In 1933 it was officially changed to Surfers Paradise at the request of local residents.

Surry Hills *NSW* is a locality in the City of Sydney. It is the subject of some disagreement. The most acceptable explanation is that it was named after Surrey in England. In 1793, Joseph Foveaux, a lieutenant in the NSW Corps, was granted 105 acres (42 hectares) of land, which he named 'Surry Hills Farm', presumably because he had come from Surrey in England. Foveaux Street in Surry Hills crosses Foveaux's original grant and thus the area gradually became known as Surry Hills. Another theory is that it is a derivation of Sunny Hills, a name which appears on Grimes Map of Sydney 1796; however, this could be a misprint as it appeared three years after the 'Surry Hills Farm'.

Sutherland *NSW* is a shire on the south-eastern outskirts of Sydney. Sutherland, according to Mrs M. Hutton Neve, research officer of the Sutherland Shire Historical Society, was the name first of a civil parish (called Southerland), then of a township and finally of the present shire. There are two theories about how they came to be named. The first is that all three are named after Forby Sutherland, the Endeavour sailor buried at Kurnell. Captain Cook named Point Sutherland in memory of him, but there is no other proof of this theory.

The second concerns the Parish of Southerland which was proclaimed in 1835. This spelling later appeared in the Government Gazette as Sutherland. The Civil Parishes were named by Sir Thomas Mitchell in 1835 when Surveyor-General. He named the first parish south of Georges River, the Parish of Southerland. Omitting the 'o' has destroyed the original significance of the name.

The Sutherland Railway Station was opened on 26 December 1885, possibly (no documented proof) named after John Sutherland, Minister for Works 1860–1872, who fought for the extension of the railway line across Georges River. The later township took its name from the railway station. The Shire of Sutherland was proclaimed in the Government Gazette of 16 May 1906 by the Governor, Sir Harry Holdsworth Rawson. According to the (Shires) Proclamation the name was selected by the Governor on the advice of the Executive Council.

Sutherland, Point *see* **Point Sutherland**

Swan Hill *V*, a city on the Murray River, was named by Major Thomas Mitchell on 21 June 1836 because of the number of black swans that disturbed his night's rest there.

Swan River *WA* was named by the Dutch explorer, Willem de Vlamingh, who came ashore near the present Cottesloe Beach on 4 January 1697. The party marched eastwards and Vlamingh gained his first view of the Swan River from Buckland Hill. The party originally mistook the river for a lake. The next day they found the black swans which resulted in Vlamingh giving the Swan River its name, Swaenerevier (River of Swans).

Swanbourne *WA* was named in 1886 by Governor Stirling, after Captain Charles Fremantle's eldest brother, the 1st Baron Cottesloe of Swanbourne, Buckinghamshire, England.

Swansea *NSW* was named by Captain R.H. Talbot, one of the pioneers of the shipbuilding trade who thought that he perceived some resemblance to the Welsh coal port.

Swansea *T* was probably named after the seaport in Glamorganshire, Wales by George Meredith, a Welshman.

Sydenham *NSW,* a suburb of Marrickville, had a station which was originally known as Marrickville Railway Station. It was later named Sydenham after a cottage in that area. Sydenham is a suburb in South-East London.

Sydenham *V* is a suburb of the City of Keilor on the northern outskirts of the Melbourne metropolitan area. Sydenham was so named because the surveyor liked the sound of the name of the place in London.

Sydney *NSW* was named by Governor Phillip after Thomas Townshend, 1st Viscount Sydney who was Secretary of State for the Colonies at the time of the founding of New

South Wales. Phillip headed his early dispatches 'from Sydney Cove' but after 17 June 1790 he wrote 'from Sydney'. Surgeon John White in his journal written in June 1788 stated that the Governor called the district Cumberland and that it was understood that the town was to be named Albion. The district around Sydney is still called the County of Cumberland but the name Albion was never adopted.

Sydney Street Names In Phillip's time George Street was a muddy track known as Spring Street and Sergeant Majors Row. The locals also called it High Street. As they exist today, the streets of central Sydney were mainly named by Governor Macquarie. He renamed George Street after King George III. He also named streets after the royal dukes of Kent, Clarence and York. Other streets were renamed to honour early governors, Hunter and Bligh. He called Macquarie Street after himself and Elizabeth Street after his wife. Martin Place is named after Sir James Martin (1820–1886), an Irish-born politician and Chief Justice of New South Wales and the Bradfield Highway over the harbour bridge honours Dr Bradfield (1876–1943) who designed the bridge. The Cahill Expressway was named after Joseph Cahill (1891–1959) who was Premier of New South Wales from 1952 to 1959.

Sylvania *NSW* in the Sutherland Shire is believed to have been named by James Murphy because of the sylvan appearance.

Symonston *ACT* is named after Sir Josiah Symon, one of the founders of the Constitution.

Syndal *V* is a suburb of Waverley in the Melbourne metropolitan area. Syndal is named after a property owned by Sir Redmond Barry, the 1852 Supreme Court judge who sentenced Ned Kelly. It was situated at the corner of High Street and Lawrence Road. 'Syndale' was farmed by Mrs Louisa Barrow, mistress of Sir Redmond Barry.

Tahlee *NSW* was the Australian Agricultural Company site headquarters and settlement. 'Tahlee House', Church, cemetery and general layout of settlement still remain very much as they used to be in the past.

Tahmoor *NSW* An Aboriginal term for the Bronzewing pigeon and the name given to his residence by James Crispe.

Tailem Bend *SA* was proclaimed in 1887. It is located on a big bend of the Murray River—hence the 'bend' in the name. The 'Tailem' has been the subject of considerable controversy. The most likely explanation is that 'tailem' is a corruption of an Aboriginal word spelt in various different ways—*thealam, thelim,* or *taileam,* depending on the recorder of the name. The name probably meant 'bend'. Mr George Mason, a subprotector of Aboriginals at Wellington, recorded it in the 1840s as Thealem. The railway junction was included in a 'run' held by D. Gollan who called it 'Taleam'. Before that it was called Pine Camp. Another explanation was advanced by Rodney Cockburn who said that overlanded cattle and sheep used to be tailed, or herded, at the bend. Another interesting but unlikely version along the same lines was advanced by A.H. Ackland who claimed that an overlander had difficulty with one of his bullocks when swimming the river. His Aboriginal boy, hoping to get the bullock moving, cried: 'Bendum tail, boss!' and the name stuck to the town.

Tallangatta *V* is a township on the Mitta Mitta arm of Lake Hume. The name comes from an Aboriginal word meaning 'many trees'. When the town was transferred in 1955-1956, more than a hundred years after its foundation on account of an extension of the Hume Weir, the Governor-General, Sir William Slim, remarked at the official opening that he approved of the retention of the name, for it sounded 'like the ring of a blacksmith's hammer'. According to the Australian Broadcasting Corporation, the correct pronunciation of this name is Ti-LANG-it-i not TAL-an-gatt-ah.

Tally Ho *V,* is a suburb of the City of Waverley in the Melbourne metropolitan area. Tally Ho takes its name from the traditional hunting call because a hunt club operated in the district at the time of Sir Redmond Barry, the judge who condemned Ned Kelly to death.

Tamar *T* is a river in northern Tasmania on which Launceston is located. In 1804, Lieutenant Colonel William Paterson landed at the present site of George Town. He explored the river, which he named in honour of Governor King, who was born at Launceston on the River Tamar in Cornwall, England.

Tamarama *NSW,* in the Sydney suburb of Waverley, was known in the 1860s as Dixon's Bay, said to be after a Dr Dixon, a nearby landowner. The first reference to Tamarama was in 1885, when a fatality was recorded at the beach. However, a map of the coastline in the 1860s by the military or naval authorities showed the name as Gamma Gamma, probably an Aboriginal name.

Tamborine *Q* comes from a corruption of an Aboriginal word that has been spelt in several ways—Dumberin, Jambooin, Tambourine and Tchambreen, all of which mean 'place or cliff of lime trees or yams'.

Tambourine Bay *NSW*, a locality in the Municipality of Lane Cove, comes from a notorious Sydney character named Tambourine Nell, who had retired there for a season to avoid the attentions of the police. After a long search they found her sylvan retreat and took charge of her and her tambourine.

Tamworth *NSW* is named after the town in Staffordshire that was represented in the British Parliament by Sir Robert Peel, the British prime minister. It was probably named in 1818 by the explorer, John Oxley, who also named the Peel River. The name means 'homestead on the River Tame'.

Tanilba *NSW* was the site of 'Tanilba House' and the interesting Halloran Development Scheme.

Tantanoola *SA* is a name whose origin is obscure. Rodney Cockburn's book *What's In a Name* says it could come from a Malay word meaning 'brushwood shelter'. *Tantanoola* is also the Aboriginal name for the peewee or magpie lark. A further possibility is that the name had something to do with the fact that Tantanoola was used as a meeting place for various Aboriginal tribes. N.B. Tindale says the name comes from Tentunola meaning 'boxwood hill' or 'boxwood camp'. Talbot says it possibly means 'hut constructed from boxwood'. The Aboriginals may have referred in this way to early settlers huts. The town was originally proclaimed in 1879 and named Lucieton. It was altered to Tantanoola in 1888.

Tanunda *SA* is an Aboriginal word signifying abundance of wild fowl in a creek.

Tarban Creek *NSW* may be named from the anglicised spelling of an Aboriginal word meaning 'fish' or 'fishing place'. However, there is no proof of this theory.

Tarcutta *NSW* An Aboriginal word meaning 'grass-seed flour'. The Tarcattah holding of 1848 was occupied by A. & G. McLeay.

Taree *NSW* comes from an Aboriginal word *tarrebit*, 'fruit of the wild fig'.

Taren Point *NSW*, in the Sutherland Shire is of unknown origin. It was at first known as Comyns Point, then Cummins

271

Point and finally as Common's Point, the origin of which is also unknown.

Taronga Park *NSW,* the site of Sydney's famous zoo, comes from an Aboriginal word for 'water view' which is appropriate as the zoo overlooks Sydney Harbour.

Tasman Peninsula *T;* **Tasman Sea** After Abel Janszoon Tasman. The name, Tasman Sea, was recommended in 1890 by the Australian and New Zealand Association for the Advancement of Science. It is usually referred to as 'the Tasman'.

Tasmania It was many years before Tasmania was proved to be an island. After sighting the coast on 24 November 1642, Tasman wrote on the following day: 'This land being the first land we have met with in the South Sea, and not known to any European nation, we have conferred on it the name "Anthony van Diemenslandt", in honour of the Hon. Governor-General, our illustrious master, who sent us to make this discovery; the islands circumjacent, so far as known to us, we have named after the Hon. Councillors of India, as may be seen from the little chart that has been made of them'. On 3 December, Tasman took formal possession of the land he had discovered. The anglicised form, Van Diemen's Land, remained in use until 1855 in official documents, but by the middle of the 1820s, the modern name, Tasmania, (in honour of the discoverer) was beginning to come into general use.

Tatura *V* is a township twenty kilometres west of Shepparton. It means 'small lagoon'.

Taylors Bay *NSW,* a locality in Mosman in Sydney's North Shore, between Chowder Head and Bradleys Head, is named after Lieutenant James Taylor, a member of the 73rd Regiment based in Sydney in 1810.

Taylors Beach *NSW* was once called Banks Farm after Captain Banks.

Tea Gardens *NSW* was a place associated with early timber and fishing industries and the birth of commercial enterprise in Port Stephens. It was named after abortive Australian Agricultural Co. attempts to grow tea there.

Tea Tree Gully *SA* is a suburb and a corporate city in the north-eastern section of the Adelaide metropolitan area. Tea Tree Gully was originally the name given to a steep gully on the north-east face of the Adelaide Hills where tea-trees once grew abundantly. The name was later given to the little town of Steventon which began to develop at the entrance to the gully in 1854, following the erection of a large wheat store by J. Stevens.

Tecoma *V* is a locality in the Shire of Sherbrooke on the outskirts of the Melbourne metropolitan area. Tecoma was the name selected for the railway station opened in 1924. Mrs J. Burke of Toorak won the prize of a guinea for the most suitable name. She was prompted by the fact that the tecoma plant flourished in the neighbourhood.

Telegraph Point *NSW* was the place where the telegraph line from Port Macquarie to Kempsey crossed the Wilson River in the early 1920s.

Telopea *NSW* was named for the waratah flower *Telopea speciosissima.*

Temora *NSW* Named after Temora Castle in a poem by Ossian. The name was given by J.D. Macanash in 1848. The township was formed at the time of the gold rush in 1880. It was proposed to call it Watsonford, but the miners insisted on retaining the name of Macanash's property.

Tempe *NSW,* a suburb of Marrickville, was named from the home of Mr A.B. Spark, now in use as a convent and in a good state of preservation on the southern bank of Cooks River, between the road and the railway bridges.

Templestowe *V* is a suburb in the City of Doncaster and Templestowe in the Melbourne metropolitan area. Templestowe is believed to have come either from the mythical village named Templestowe, 'place of temples', in Sir Walter Scott's novel *Ivanhoe,* or a small settlement in England around 1765 known as Templestowe. There is nothing to show who gave the name or why it was given.

Tennant Creek *NT* was named by John McDouall Stuart when he passed through the region in 1860. He called it after a friend named John Tennant of Port Lincoln.

Tennyson *SA* is a suburb of Woodville, a corporate city in the Adelaide metropolitan area. It was named after the Governor of South Australia, Hallam Tennyson, 2nd Baron Tennyson (and a son of the famous English poet). It was proclaimed in 1905.

Tenterfield *NSW is* named after the family estate of Stuart Alexander Donaldson, the first premier of New South Wales. He gave the name to his New South Wales property in the 1840s. This was taken for the town that was gazetted in 1851.

Terang *V* is a township forty-five kilometres north-east of

Warrnambool. Its name is Aboriginal for 'bare twig'.

Terrey Hills *NSW,* a locality in Warringah Shire, was taken from a trig station there named Terrey. The trig station was called after a pioneer named Terrey, who was granted land there. 'Hills' could be because it is hilly, or after the Hills, an old pioneering family that still live there.

Terrigal *NSW* is an Aboriginal word. It may mean 'place of little birds'.

Texas *Q* is a town near the Dumaresq River on the New South Wales border, about 270 kilometres west of Brisbane. The area was first settled in 1842. The town's name probably comes from the fact that the Texas district, like its American namesake, was disputed country. Malcolm Septimus McDougall took up the land before joining his brothers on the Turon goldfields. Returning to their neglected station, the brothers found the country had been 'jumped' by another enterprising settler, and it took some considerable time before they were able to establish their prior claim.

Thargomindah *Q* is an Aboriginal word for 'spiny anteater'.

Tharwa *ACT* is a name of Aboriginal origin associated with the district since the early days of the settlement.

The Basin *V* is a locality in the City of Knox on the outskirts of the Melbourne metropolitan area. The Basin was named by the famous botanist Baron von Mueller because of the basin formed by the surrounding hills. He first visited the district in 1853 and during the 1870s established a permanent camp there.

The Patch *V* is a locality in the Shire of Sherbrooke on the outskirts of the Melbourne metropolitan area. The Patch was named by Isaac Simmons, a timber worker in the 1880s who found a patch of blackbutts suitable for palings growing along the site of the post office.

The Rocks *NSW* is a locality in the City of Sydney. The Rocks is an area of special historical interest and one of Sydney's major tourist attractions. It is a rocky sandstone outcrop adjoining Sydney Cove.

The Spit *NSW,* a locality in Mosman on Sydney's North Shore, gets its name from the sand spit which runs out into Middle Harbour at this point. The road from the city to Manly crosses Spit Bridge which can be raised and lowered to allow ships to pass through the narrow channel.

Thebarton *SA* is an inner suburban area of Adelaide. In 1838 Thebarton and other areas of Adelaide were surveyed by

Surveyor-General Colonel William Light and his team of assistants. Light won a ballot to select land and chose an area along the River Torrens. A house was built for Light on the corner of what are now Cawthorne and Winwood Streets. He named the house 'Theberton Cottage', after 'Theberton Hall' in Suffolk, England, where he spent part of his youth. The incorrect spelling of Theberton to Thebarton may have been a typographical error. The preface to *A Brief Journal of the Proceedings of William Light,* published in 1839 is dated from 'Thebarton Cottage', near Adelaide. Thebarton was the first village laid out beyond Adelaide and was proclaimed a corporate town on 7 February 1883. It was surveyed in July 1839, for Colonel Light by Nixon and Finniss.

Theodore *ACT* is called after Edward Granville Theodore, who was Premier of Queensland from 1919 to 1925, and a member of federal parliament from 1927 to 1931.

Theodore *Q* is named after E.G. Theodore, who was Premier of Queensland from 1919 to 1925 and Federal Treasurer in 1929.

Thevenard *SA* was proclaimed a town in 1924 and took its name from Cape Thevenard. Nicolas Baudin originally called this cape, Bon Fond, 'good anchorage'. However, he died before he could return to France. The cartographer on the voyage, Lieutenant Freycinet, later published charts on which he changed many of Baudin's original names, as Baudin's original charts had not been published at the time. Freycinet named Cape Thevenard to honour Antoine Jean Marie Thevenard, a French admiral and minister of marine. Flinders, who adopted the French names along the coast east of Encounter Bay did not apply a French or English name to this cape. The name Cape Thevenard did not come into general use until after the settlement of South Australia.

Thirlmere *NSW* is named after a lake in Cumbria.

Thirroul *NSW* is Aboriginal for 'valley' or 'hollow'.

Thirsty Sound *Q* 'This inlet,' wrote Cook on 30 May 1770, 'which I have named Thirsty Sound, by reason we could find no fresh water . . . '

Thistle Island *SA* lies at the western entrance to Spencer Gulf SA. The British explorer Matthew Flinders discovered the island in 1802 during his survey of the southern shores of the continent in the *Investigator*. Flinders named it Thistle's Island after the master of the *Investigator,* John Thistle, who drowned with seven other members of the crew when their cutter capsized while they searched for an anchorage and water supply.

Thomson *V* is part of Geelong. It is named after Dr Alexander Thomson, one of Geelong's first settlers in 1836, first elected Mayor of Geelong, and politician.

Thomson River *Q* was named in 1847 by the explorer, Edward Kennedy. He named it after a Sydney University chancellor, Sir Edward Deas Thomson.

Thornbury *V* is a suburb of the City of Northcote in the Melbourne metropolitan area. Thornbury comes from Thornbury Park formed on a section of land by a subdivision of Northcote. The original land was part of the estate of Job Smith on Merri Creek. The English Thornbury is in Hertfordshire.

Thorndon Park Reservoir *SA* is part of the suburb of Paradise in the City of Campbelltown in the Adelaide metropolitan area. The reservoir was built in 1857 as the first public works on a large scale to provide a water supply to Adelaide. Land in this area was granted to Lord Petre and the Honourable Henry Petre of Thorndon Hall (near Brentwood in Essex, England) in 1838—a likely explanation of the origin of the name. Lord William Henry Francis Petre was a member of the New Zealand Association which paved the way for the imperior colonization of New Zealand. The reservoir is on land repurchased partly from the Petres and partly from the South Australian Company. The small private subdivision of Thorndon Park is now included in the suburb of Newton.

Thorngate *SA* is a suburb in the corporate city of Thorngate, just north of the centre of Adelaide. It was granted to John Battley Thorngate of Gosport, Hampshire, England, in 1840. He died in 1867 and left it to his brother William who, in turn, died in 1868. In 1867 William had conveyed the land to E. Churcher and others to be held in trust. The annuities from the land were to be paid to numerous charities. In 1882 part of the land was subdivided into sixtynine allotments and the area was called Fitzroy. In 1911 the trustees Geo. and W.E. Churcher had the Thorngate Estate (Private) Act passed, vesting the land to them, freeing the land from the trust and securing the annuities. In 1913 they had the land subdivided into fifty-nine allotments and the area was named Thorngate.

Thornleigh *NSW*, a locality in Hornsby Shire on the North Shore of Sydney, is named after Chief Constable John Thorn (1794–1838) who was given a grant of land in the area for his part in the capture of an armed bushranger named McNamara, on the Windsor Road.

Thornlie *WA,* in Gosnells city in the Perth metropolitan area, derives its name from 'Thornlie Homestead'. This was the name chosen by Frank James (1861–1929), a young farmer from the Yatheroo district, when he brought his new bride to the newly built homestead in 1884. He chose the name because his grandfather was in business in Thornlie Bank Madras, in India. The land was originally selected in November 1829 by Captain Thomas Bannister who named his property 'Woolcombs'. It was then owned by Walter Padbury and run as an experimental farm by his niece's husband, Frank James. In December 1955 the Gosnells Road Board requested a new subdivision to be named Thornlie, which appeared to be a name by which the whole estate was known in the early days. The name was approved on 12 April 1956.

Thredbo *NSW* This is an Aboriginal name.

Thursday Island *Q,* in the Torres Strait about forty kilometres north of Cape York, is the administrative centre for the Torres Strait Islands. It is not known who named Thursday Island. The names 'Thursday' and 'Friday' Islands first appear in Captain Owen Stanley's survey data drawn up after his 1848 voyage in the *Rattlesnake,* but in the reverse order from what they are today.

Tibooburra *NSW* is an Aboriginal word meaning 'rocks' or 'heaps of boulders'.

Tilba Tilba *NSW* is Aboriginal for 'many waters'.

Tin Can Bay *Q* is located near Gympie north of Brisbane. The name is derived from the native word *tindhin,* the Aboriginal name for a species of mangrove.

Tinaroo Creek *Q* is said to have got its name because John Atherton cried 'Tin, Hurroo' when he made the discovery of tin there.

Tocumwal *NSW* An Aboriginal term meaning 'deep hole'. The local Aborigines believed that the hole in the river extended from here to The Rock, eleven kilometres to the north-east.

Tod River *SA* is named after its discoverer, Robert Tod, who was employed by the South Australian Company in 1839. It is the only river of any size over a range of 1600 kilometres from Albany to Port Augusta. In his book, *What's In A Name,* Rodney Cockburn says, 'The stream takes its rise in a beautiful valley, called by Tod, Cowan Vale without recording the name'. According to *The South Australian Almanack* of 1840, it was discovered whilst exploring Spencer

Gulf in the schooner *Victoria* by a party consisting of Messrs Austin, Crouch, Finn and others. In the course of their exploration trip into the interior, they discovered the river and named it after Robert Tod, one of their party.

Todd River *NT* A survey party from the Overland Telegraph Line discovered the river and named it after the Postmaster-General, Sir Charles Todd, who was responsible for the construction of the line. The dry bed of the stream, which loses itself in the Simpson Desert at Alice Springs, is the scene of the famous annual 'boat race', in which the 'boats' are carried by runners. At certain seasons the river becomes a rushing torrent.

Tom Uglys Point *NSW* in the Municipality of Kogarah is the subject of two theories. The first, which is probably correct, is that one of the early residents on the north side of Georges River was a Tom Huxley. The Aborigines could not pronounce the 'X' and the surname gradually became corrupted to Ugly. The second is that the name derives from a white man with only one leg and one arm living in the locality known to the natives as Tom Wogully, or Wogul. *Wogul* means 'one', hence the Aboriginal name for the settler Tom Wogully was corrupted to Tom Ugly.

Tomah, Mount *see* **Mount Tomah**

Toodyay *WA* was originally known as Newcastle and it was gazetted as such on 9 October 1860. The name was changed to Toodyay in the gazette of 6 May 1910. It was named after the nearby brook recorded as Toodyay by George Fletcher Moore in 1836 on Exploration Plan 126. Moore and Ensign Dale crossed this brook in 1831. It is recorded that Toodyeep was the pretty wife of the Aboriginal, Coondebung, who accompanied Moore and Dale in 1831. The name is pronounced 'Toojy'.

Toolern Vale *V* is a small township eight kilometres north of Melton. The Aboriginal word *toolern* meant 'tongue'. The school was originally named Yangardook. It was changed in 1875 to Toolern, and renamed Toolern Vale in 1931.

Toongabbie *NSW*, a locality in the City of Blacktown and the Municipality of Holroyd on the western outskirts of Sydney, is an Aboriginal name for 'meeting of waters'.

Toorak *V*, one of Melbourne's best known up-market suburbs, is located in the City of Malvern. Toorak takes its name from an estate bought by a merchant, James Jackson, in 1843. He built a mansion there in 1849 and named it 'Toorak House'. The Aboriginal word meant 'swamp where tea-trees grow'.

Toowong *Q* is a locality in the Brisbane metropolitan area and gets its name from the Aboriginal interpretation of the sound of the goatsucker bird.

Toowoomba *Q* is a city about 130 kilometres west of Brisbane. In 1840 the Leslie Brothers established 'Toolburra' station near Warwick. Other settlers were quick to follow. A town of sorts sprang up nearby called The Springs. One of the early settlers at The Springs was Thomas Alford, who arrived in 1842 with his wife and two daughters. He did not like the name of the settlement and renamed it Drayton, after a place in Somerset, England. Drayton today is a suburb on the main highway south and the main city is named after 'Toowoomba', Alford's house.

The origin of the word Toowoomba is uncertain. It may have come from 'Tchwampa', an Aboriginal corruption of 'The Swamp', which was the original name for the village that grew up in a marshy area in the 1850s. Another explanation gives the name as coming from *toowoon* or *choowoon,* meaning the native melon that grows in the swamp. It could also have come from *toowoomba* meaning 'underground water'.

Toronto *NSW* was named after a city in Canada by the Excelsior Investment and Building Company in honour of Edward Hanlan who was then a champion sculler. He arrived in New South Wales from Toronto at the time the company's land was being subdivided.

Torquay *V* is located twenty-one kilometres south of Geelong. Like its English namesake, it is a seaside holiday resort. The Victorian resort has a famous surf beach which has been the venue for a number of international and Australian surfboard championships.

Torrens *ACT* honours Sir Robert Torrens (1814-1884) who, as Premier of South Australia in 1857, introduced the Torrens system of land title now used throughout the world.

Torrens, Lake *see* **Lake Torrens**

Torrens Park *SA* is a suburb in the corporate city of Mitcham in the Adelaide metropolitan area. It takes its name from 'Torrens Park House', which is listed as part of the National Heritage. It was built in 1854 for Sir Robert Torrens, the originator of the land titles system. Part of it was formerly called 'Blythwood' after Sir Arthur Blyth, who lived in Mitcham for many years in a house called 'Whitehall'.

Torrens River *SA* was discovered by Lieutenant W.G. Field, John Morphett and G.S. Kingston in November 1836, but named by Surveyor-General William Light, after Colonel Robert Torrens, chairman of the South Australia Colonization Commission and father of Sir Robert Richard Torrens who became Premier of South Australia in 1857. The river has various Aboriginal names depending on where it flows. The portion in the city was called Karra wirraparri and Korra-weera meaning 'river of the red gum forest'. At Hindmarsh it was called Karraundo-ingga; at Reedbeds, Witoingga. Throughout its whole length, it was called Perre, Peere or Parri. In flood, it was called Yertala.

Torres Strait *Q* was named by the British cartographer, Alexander Dalrymple, in 1767 in honour of the Spanish navigator, Luis Vaez de Torres, who sailed through the strait in 1606.

Townsville *Q* is named after Captain Robert Towns, a wealthy Sydney businessman who was the settlement's financial backer. Towns visited the settlement only once, in 1866, and then only briefly but he deserves full credit for the backing he gave to his junior partner, John Black, who acted as town planner and administrator.

Trafalgar *V*, a township in Central Gippsland ten kilometres west of Moe, was named after the famous naval battle. At one time, nearby Yarragon was called Waterloo.

Trangie *NSW* is an Aboriginal word said to mean 'quick intercourse'.

Tranmere *SA* is a suburb of the corporate city of Campbelltown in the Adelaide metropolitan area. Tranmere, meaning 'across the sea', was named by David Wylie, an early settler, after Tranmere, Birkenhead, England, where he had lived. It was one of the earliest place names in the colony. Wylie, a graduate of Glasgow University, arrived in South Australia in 1838 on the ship *Canton*. He conducted an educational establishment named 'Tranmere House' on his land.

Traralgon *V*, a city in Gippsland between Morwell and Sale, was first settled by Europeans in the 1840s. The town's name may have been derived from an Aboriginal word for 'river of little fish' or for a water bird, probably a heron. The Muk-thang Aboriginals are thought to have occupied the area before European settlement.

Tregear *NSW*, a locality in the City of Blacktown on the western outskirts of Sydney, is named after the Lethbridge family seat in Cornwall, England.

Tremont *V* is a locality in the Shire of Sherbrooke on the outskirts of the Melbourne metropolitan area. Tremont was named at the suggestion of F.J. Treweek when a post office was established there in 1918. The name literally means 'mountain of trees'.

Trephina Gorge *NT* is called after a pioneer named Trephina Benstead.

Triabunna *T* is of uncertain derivation It was first called Port Monthazin by Nicolas Baudin after one of his men, and then Spring Bay, so called by George Meredith after his dog.

Trial Bay *NSW* takes its name from the brig *Trial* which was wrecked there in 1816. A search party led by Commander White discovered the wreck at the mouth of the Macleay River in 1817.

Tribulation, Cape *see* **Cape Tribulation**

Trinity Bay *Q* was named by James Cook because it was discovered on the eve of Trinity Sunday.

Truro *SA* is named after Truro in Cornwall, meaning 'village on a hill'. According to Rodney Cockburn's *What's In A Name,* the village was named by John Howard Angas (son of George Fife Angas) in 1847–48. It means 'three roads' in Old English.

Tuart Hill *WA*, a suburb of Stirling, a city in the Perth metropolitan area, derived its name from the fine old tuart trees on the high land on either side of Wanneroo Road when the land was subdivided for sale in 1905. Unfortunately, most of these trees disappeared when more houses were built after World War II.

Tuggerah *NSW is* an Aboriginal word meaning 'cold'.

Tullamarine *V* is the site of Melbourne's airport. Tullamarine was named after a small boy called Tullamareena, a member of the Warundjeri tribe who was noticed in 1837 by the missionary, Samuel Longhorne, who spoke about him to the surveyor, Robert Hoddle.

Tully *Q* was named in 1872 after the Surveyor-General of Queensland at the time, William Alcock Tully.

Tumbarumba *NSW* An Aboriginal name meaning 'sounding ground'. The name is onomatopaeic, for the ground here gives a hollow sound when stamped on. An early pastoral holding here was named 'Tombrumba Creek'.

Tumble Down Dick *NSW* is a locality on the Mona Vale Road in Warringah Shire. According to local legend, this intriguing name commemorates a blind horse named Dick, which fell down the hillside to its death while leading a bullock team. A more likely explanation is that it takes its name from Tumbledown Dick, a popular name for hotels in England, which in turn got their name from a derogatory reference to the drunken son of Oliver Cromwell.

Tumut *NSW* comes from an Aboriginal word *doomat* or *doomut* meaning 'camping place by the river'. The explorers Hamilton Hume and William Hovell passed through the area in 1824. Thomas McAlister became the first settler in the Tumut Valley when he founded 'Darbalara' station in the 1830s. A town-site was surveyed in 1848. A municipal council was formed in 1887.

Tuncurry *NSW* means 'honey'.

Turner *ACT* is called after Sir George Turner, a founder of the Constitution.

Turrella *NSW*, a suburb of the Municipality of Rockdale, is named after an original land grant.

Tweed Heads; Tweed River *NSW*, *Q* Named by John Oxley in 1823 after the Tweed River which, in part, divides England from Scotland. In 1828, Captain Rous came upon Oxley's discovery and, believing himself to be the discoverer, named it the Clarence. When the mistake was revealed, the name Clarence was transferred to what was then known as Big River.

Tyrrell, Lake *see* **Lake Tyrrell**

Uhrs Point *NSW* is a locality in the Sydney municipality of Concord. It was named in honour of George Richard Uhr, Sheriff of Sydney, who built a prominent house there in the 1840s.

Uley *SA* is the name of two places in the state of South Australia.

1. Uley in the Munno Para district north of Adelaide got its name because the hills of the locality reminded Moses Bendle Garlick, one of the area's first settlers, of his home village of Uley in Gloucestershire, which is dominated by The Bury, a hill containing ancient Britons' graves.

2. Uley is also an area on the Eyre Peninsula. The country was under pastoralists E. Spicer, R. Symes and J. Sinclair in 1850. Uley was proclaimed a 'hundred' in 1871.

Ulladulla *NSW* comes from an Aboriginal word said to mean 'safe harbour'.

Ultimo *NSW* is a locality in the City of Sydney. Ultimo got its name because of a slip of the pen. In 1803, Surgeon John Harris was court-martialled for allegedly disclosing information given during the court martial of a fellow officer. The charge referred to the date of the offence as the 19th ultimo instead of the 19th instant and, on this technicality, Harris was acquitted. In the following year he received a grant of land from Governor King and on it he built an imposing mansion which he appropriately named 'Ultimo House'. The building was demolished in 1932 and its deer park is now the site of the Sydney Morning Herald's printing building.

Ulverstone *T* is a municipality on the banks of the Leven River. Nobody knows who named Ulverstone, but it was probably someone familiar with Ulverston in England and knew it was on the Leven River.

Unley *SA* is a corporate city just south of central Adelaide. Unley was named by Thomas Whistler, who arrived in 1840. When he subdivided his land he called the village Unley. This was originally thought to be after Whistler's wife's maiden name, but it is more probably called after Undley, a hamlet in the parish of Mildenhall and Lakenheath in Suffolk, his home county. The 'd' seems to have been dropped.

Upwey *V* is a locality in the Shire of Sherbrooke on the outskirts of the Melbourne metropolitan area. Upwey was named by Sarah Tullidge, a retired St Kilda schoolmistress who settled in the area in the 1880s. The Tullidge family came from the English village of Upwey by the River Wey.

Urangan *Q* means 'small white shells' in the local Kabi Kabi language.

Urango *NSW* is an Aboriginal term for 'long beach'.

Vale Park *SA* is a suburb within the corporate towns of Walkerville and Enfield in the inner metropolitan area of Adelaide. Vale Park takes its name from 'Vale House' which, in the 1860s, under the ownership of Philip Levi, a pastoral pioneer and prominent Adelaide businessman, was the centre of the Adelaide Hunt Club. Philip Levi arrived in

South Australia in 1838 aged sixteen years and worked with sheep, firstly as a drover and then as the owner of sheep runs north of Adelaide. The homestead is now part of 'Levi Park', a caravan park in Vale Park.

Valley Heights *NSW,* a locality in the Blue Mountains west of Sydney, got its name because in 1831 an inn, Valley Inn, was erected to serve travellers at Fitzgerald's Valley. By 1855 the locality was called The Valley and this name was used when the railway station was built in 1875. The name was changed to Valley Heights in 1890.

Vaucluse *NSW* is located in the Municipality of Woollahra. Rightly or wrongly it is generally considered to be one of Sydney's wealthiest suburbs. Vaucluse is a French form of the Latin *vallis clausa,* 'enclosed valley'. The name probably came to Sydney by way of the name of a town in south-east France made famous in Italian literature by the poet Petrarch. It was chosen by Sir Henry Brown Hayes, a wealthy Irish convict, to apply to his land where 'Vaucluse House' now stands.

Victor Harbor *SA* was named after HMS *Victor* by Captain Richard Crozier who sailed in this vessel when he surveyed the harbour in 1837. The name is officially spelt Victor Harbor using the American spelling of 'harbour' because Captain Crozier's maps and log used this form. In fact, all named harbours in South Australia use the spelling 'harbor', possibly because of the existence of the Harbors Act, the now redundant Harbors Board and the present Department of Marine and Harbors.

Victoria was named after Queen Victoria when the colony separated from New South Wales in 1851. Prior to this, the area was usually referred to as the Port Phillip district.

Victoria, Lake *see* **Lake Victoria**

Victoria, Mount *see* **Mount Victoria**

Victoria, Port *see* **Port Victoria**

Victoria River *NT* was discovered and named in honour of Queen Victoria by Captain J.C. Wickham in 1839.

Villawood *NSW* is a suburb of Fairfield on the south-western outskirts of Sydney's metropolitan area. It has a very interesting origin. Like Carramar, the name was decided by the Railway Department when the Regents Park to Cabramatta line was opened in 1924. However, the original choice of name was Woodville (after the historic road which runs by the station), but when research showed that another town named Woodville existed near Newcastle, it was decided to change the name, but in order to retain the local historic association with the Woodville Road, the two syllables Wood and Ville were reversed and rounded into Villawood.

Vineyard *NSW*, a locality in the City of Blacktown on the western outskirts of Sydney, was originally part of Windsor's winegrowing districts.

Violet Town *V*, a township twenty-four kilometres south-west of Benalla, was originally named Violet Ponds by Major T.L. Mitchell in 1835 because of the violets he found growing by a chain of ponds. In 1852 an inn by the Honeysuckle Creek marked the site of the proposed settlement of Violet Dale.

Virginia *SA* was surveyed in 1858 for the proprietor, a wealthy Irishman, Daniel Brady, who called it after Virginia in County Cavan, Ireland which, in turn, is named after the Virgin Queen. It was originally an Irish settlement of people who had fled the potato famine in Ireland.

Wacol *Q* is a locality in the Brisbane metropolitan area and was named by the Railways Department in 1927. Wacol is an abbreviation of the words 'weigh' and 'cool'.

Wagga Wagga *NSW* is an Aboriginal phrase meaning 'place of crows'.

Waikerie *SA* was named by W.T. Shephard who set up a station there in 1882. According to Shephard the name came from an Aboriginal word. It may mean 'anything that flies' or 'favourite site for waterfowl to settle'. The anthropologist N.B. Tindale states that it is derived from *weikari,* the name given to a species of ghost moth, which, at a certain season each year, appeared in thousands among the river red gums, at which time they provided valuable food for the Aboriginals. The word also applies to the actual flight of the moths. The township was proclaimed by Governor Bosanquet in 1910.

Wait-A-While *NSW* comes from the name of a holding of R.T. Kister.

Waitara *NSW,* a locality in Hornsby on the North Shore of Sydney, is named after a town in Taranaki in New Zealand. The Maori name may mean 'mountain stream'.

Wakeley *NSW* is a suburb of Fairfield on the south-western outskirts of the Sydney metropolitan area. It is named after a family who owned one of the first farms in the area.

Walcha *NSW* An Aboriginal term for the sun. The name was taken from a pastoral holding.

Waldheim *T* is German for 'forest home'. It was built by Gustav Weindorfer (1873-1932), the founder of Cradle Mountain Park.

Walgett *NSW* comes from an Aboriginal word with a number of possible derivations. They include 'long waterhole', 'plenty of water', 'swamp', and 'high hill'.

Walhalla *V* is an old goldmining town which was first called Stringer's Creek, after Ned Stringer, a prospector who first discovered gold there in 1862.

Walkerville *SA* is an inner suburb of the Adelaide metropolitan area. Walkerville was named after Captain John Walker, RN who called a public meeting in 1838 to plan a new township. John Walker soon ran into financial trouble and departed for Tasmania.

The original section was owned by Governor Hindmarsh, who sold to a syndicate for £1000. Meetings were held in 'Lawe's Tavern' (The Walkers Arms Hotel), which was demolished and rebuilt on the original site in the early 1970s. It was one of the oldest hotels in South Australia. It was known as the Walker's Arms as early as 1839.

Wallacia *NSW*, a locality within the boundaries of the City of Penrith, derives its name from that of a nineteenth century resident of the district, Robert Wallace, whose home became a convenient depot for local people to pick up their mail. He lived where the Wallacia post office now stands. In 1906, at the instigation of a local schoolmaster, the name was officially made Wallacia.

Wallaroo *SA* is an Aboriginal word for 'black mountain wallaby' in one of the New South Wales dialects. However, according to Rodney Cockburn's book *What's In A Name*, the South Australian name is a corruption of *wadla-waru*, which means 'wallaby's urine'. The first run established there was 'Wallawaroo' held by Captain (Sir) W.W. Hughes. The name was shortened to 'Wallaroo' because it saved time when branding wool bales.

Wallerawang *NSW* An Aboriginal word meaning 'plenty of water'. From the estate of James Walker.

Wandana Heights *V*, a suburb of Geelong, was subdivided in 1965. Mr G. Eaton was the member of the group of subdividers who decided upon Wandana as the name of the subdivision. The other members of the group decided to add Heights to the name. *Wandana* is an Aboriginal word which means 'to see far away'. The land in the area was supposedly given to William Robertson by William Buckley in about 1842.

Wandoan *Q* was originally called Juandah after the 'Juandah' station which was selected in 1853. In 1927 the name of Juandah was changed to Wandoan because it was confused with Jundah.

Wangaratta *V*, a city on the Ovens River, stands close to the site chosen by its first settler, George Faithfull. He named his property after two Aboriginal words meaning 'where the cormorants sit'. A village grew near his homestead which was close to a convenient coach and stock crossing on the river. Wangaratta became a borough in 1863 and a city on 15 April 1959.

Wangi Wangi *NSW* means 'many night owls'. It is pronounced wonjy.

Wanneroo *WA* is a shire sixteen kilometres north of Perth. *Wonna* or *wanna* means 'a stick' used by Aboriginal women for digging roots, and *roo*, it is presumed, is used to denote 'place of'. Thus Wanneroo is 'place where native women found a stick for digging roots', and as early as 1842, Surveyor P.S. Chauncey recorded a 'road to Wanneroo' and in 1844 James Dobbins gave his address as Wanneroo. It was gazetted on 16 August 1907 as Wanneru. The spelling was amended and gazetted on 15 May 1953 as Wanneroo.

Wanniassa *ACT* is a name that has been associated with the Tuggeranong district of the A.C.T. since the early days of the settlement.

Waramanga *ACT* is the name of a Central Australian tribe.

Waratah *NSW*, a Newcastle suburb, is probably named after the waratah flower.

Warburton *V* is a township seventy-four kilometres east of Melbourne by the Yarra River. When the town was established in 1864, at the time of the gold rush, it was named after Charles Warburton Carr, the district police magistrate.

Warialda *NSW* An Aboriginal name meaning 'place of wild honey'.

Warners Bay *NSW*, a locality on Lake Macquarie, is named after Jonathan Warner, the original land grantee.

Warning, Mount *see* **Mount Warning**

Warracknabeal *V* is a township fifty-seven kilometres north of Horsham. E.J. Eyre was there in 1884. In 1885, the Scott brothers established the 'Warracknabeal' station. The name in 1878 was spelt Werracknebeal. The name is said to be Aboriginal for 'large gumtrees'.

Warragul *V* is a township 103 kilometres south-east of Melbourne. The settlement was first known as Brandy Creek. The name comes from an Aboriginal word meaning 'wild'.

Warrandyte *V* is a suburb of the City of Doncaster and Templestowe in the Melbourne metropolitan area. Warrandyte originated from an Aboriginal word *warren*, 'to throw' and *dyte*, 'the object aimed at'. The area was used by Aboriginal tribes as a meeting place to hold games of boomerang and spear throwing. Louis John Mitchell discovered gold in 1865 in Anderson's Creek and diggings developed. The area was known as Anderson's Creek until 1908 when the old station name was adopted.

Warrane *T*, in Hobart's eastern shore suburb of Montagu Bay, was known as Warrane for postal purposes in the 1920s. In 1941, the name was applied to its present location, that of a Housing Department subdivision about two kilometres to the north-east. It was proclaimed as a town in 1953, but was not officially named by the Nomenclature Board until 1960. It is of Tasmanian Aboriginal origin, meaning 'blue sky', the approximate pronunciation being 'Wor-ar-nay'; now it is pronounced much more mundanely 'W-rain'.

Warrawarrapiraliliullamalulacoupalynya *NT* has been claimed to be the longest place name in Australia. However the Northern Territory Department of Lands Place Names Committee has investigated the name and can find no evidence supporting its existence. Reverend John Flynn of the Australian Inland Mission collected this name in the 1930s. Its meaning is unknown.

Warrego Range; Warrego River *Q* Both the river and the range were explored by T.L. Mitchell in 1845 and 1846, the Aboriginal name being retained.

Warren *NSW* An Aboriginal term meaning 'strong', or 'large root'. The latter meaning has a physiological significance.

Warriewood *NSW* is a suburb of Warringah Shire in Sydney's northern beaches area. Warriewood was named after

Warriewood estate which was sold by Henry Halloran, surveyor and estate agent to the McPhersons. It is not known who named it Warriewood.

Warrigal *NSW Q V* is an Aboriginal name for 'dingo' .

Warrimoo *NSW,* a locality in the Blue Mountains west of Sydney, was formerly referred to as Karabar where a railway platform and waiting shed existed in 1881. Lack of patronage caused the closure of the platform in 1897. In 1919, Arthur Rickard opened up estates in the area and the name Warrimoo was adopted, this being an Aboriginal word for 'eagle'.

Warringah *NSW* is a shire occupying most of the peninsula that stretches from Manly to Palm Beach. Warringah is an Aboriginal word. It is said to mean 'sign of rain'.

Warrnambool *V* is a coastal city in south-west Victoria. It was surveyed by Robert Hoddle and Lieutenant Pickering in 1846. The name adopted by Lieutenant Pickering is taken from a hill by the Hopkins River with an Aboriginal name, Warnimble, which is said to mean 'place of plenty'. It has also been suggested that it means 'place between two waters', the two waters being the Hopkins River on the eastern edge of the city and the Merri River encompassing Warrnambool to the north and west. Another suggested meaning is 'running swamps'.

Warrumbungle Range *NSW* is believed to be an Aboriginal name meaning 'little or broken mountains'. John Oxley explored the mountains in 1818 and called them the Arbuthnot Range, but this name did not survive.

Warwick *Q* was selected as a town-site by Patrick Leslie in 1847 and the first allotments were sold in 1850. It is named after Warwick in England.

Warwick Farm *NSW* was named by J.H. Stroud, superintendent of the Liverpool Orphan's School, after a town about fourteen kilometres south-east of Coventry in England.

Waste Point *NSW* is a location near the road leading to Perisher Valley. A large round concrete tower built by the Snowy Mountains Authority just near the road suggests that the 'waste' refers to some discharge by the SMA. But the name pre-dates the Snowy Mountains Scheme. It comes from the time when surveyors from the Lands Department were employed checking and measuring land taken up by James Spencer. It was then known as the Triangle. When the Surveyors came to the junction of the two rivers they found a triangular piece of land which is low lying and had been formed by silt deposits brought down by the rivers in floodtime. The surveyors remarked on this and said 'we will not include this in our measurement, it is waste'. Afterwards, the whole area was known as Waste Point. The homestead became the first boarding house in the Kosciusko region and it was here that Banjo Paterson stayed and wrote his famous poem, *The Man from Snowy River*.

Waterfall *NSW*, in the Sutherland Shire, derives its name from the waterfalls near the railway station. McKell Avenue, the southern boundary of Royal National Park, was originally Waterfall Road.

Waterford *Q* was named after Waterford in Ireland by the Irish settlers, who first came there about 1849. The Aboriginal name was Tygum, meaning 'large'.

Waterloo *NSW*, a Sydney suburb, recalls the Battle of Waterloo in 1815 in which the British defeated Napoleon. In 1813, John Hutchinson requested assistance from Macquarie to erect a water mill—believed to be near the junction of Bourke and Elizabeth Streets, Waterloo. In 1820 a water flour mill about three kilometres from town was erected by Messrs William Hutchinson, D. Cooper, George Williams and William Leverton. After a visit by the Governor, the name of Waterloo Mills was bestowed upon it.

Watson *ACT* honours John Christian Watson (1867–1941), the first Labor prime minister of Australia in 1904.

Watsonia *V* is a suburb of the Shire of Diamond Valley on the outskirts of the Melbourne metropolitan area. Watsonia was named after Frank Watson who bought land there about 1900. The Heidelberg Council called the railway station after him because of his efforts to raise money from local residents to pay for the station. Watson, together with Jack Black and Wilkie Joules, built the station in 1924 at a cost of three hundred pounds hoping to attract settlers to the new district.

Watsons Bay *NSW* is located on Sydney Harbour in the Municipality of Woollahra. Watsons Bay came into use officially when Robert Watson was appointed harbourmaster in 1811. He had lived there for some years previously as a pilot and Governor Macquarie adopted the name that had been used unofficially for that anchorage and the village on its shore.

Wattamolla *NSW* in the Sutherland Shire was named as a civil parish in 1835. It is an Aboriginal word meaning 'place near running water'.

Wattle Park *V* is the name of a recreation reserve in the City of Box Hill where wattles grow in abundance.

Wauchope *NSW* is named after Captain Robert Andrew Wauch who took up a grant there in 1841.

Wavell Heights *Q* is a locality in the Brisbane metropolitan area and is named after General Sir Archibald Wavell, Commander-in-Chief in the Middle East where Australian soldiers served in World War II.

Waverley *NSW*, the suburb that gives the Sydney municipality its name, got its own name from 'Waverley House', built by Barnett Levey, the entrepreneur who founded the Theatre Royal, Australia's first permanent playhouse. Levey named his mansion after Sir Walter Scott's Waverley novels, which were published in 1814. Scott in turn drew the name from the Abbey of Waverley near Farnham in Surrey, England.

Waverley *V* is a suburban and industrial city about twenty kilometres south-east of central Melbourne. The city was probably named by Dr James Silverman who made the first subdivision of land in the district in 1854 and called it the Waverley Estate. The Oakleigh and Mulgrave Roads District was created in 1857. Oakleigh became a shire in 1871. Waverley became a city in 1971.

Waverley, Mount *see* **Mount Waverley**

Waverton *NSW*, a locality in North Sydney, is named after a house bought by William Carr, a solicitor, in 1850 and named by him after a village in England. Richard Old, director of the Australian Gas Light Company, and his descendants, owned the house from 1865 to 1974. The modern townhouses there still use the name Waverton.

Wedderburn *V* is a township by Korong Creek, seventy-four kilometres northwest of Bendigo. Gold was discovered there in May 1852 and the field is known as Mount Korong. Wedderburn is named after a stream in Scotland.

Wee Waa *NSW* comes from an Aboriginal word meaning 'fire thrown away'. The first cattle station there was established in the 1840s by two settlers named Campbell and Ryan, who called their property 'Weeawaa'.

Weetangera *ACT* has been a name associated with the locality since the days of the early settlers.

Wellington *NSW* takes its name from the Wellington Valley which was named by John Oxley in August 1817 during the search for the Macquarie River. He probably called it after the Duke of Wellington (1769-1852), the hero of Waterloo.

Wellington, Lake *see* **Lake Wellington**

Wellington, Mount *see* **Mount Wellington**

Welshpool *WA* is derived from Welshpool Road which was known by this name at least as early as 1889. The suburb name was first recorded in 1895 in the *Western Australian Post Office Directory,* as a railway siding.

Wendouree, Lake *see* **Lake Wendouree**

Wentworth Falls *NSW,* a locality in the Blue Mountains west of Sydney, was named after William Charles Wentworth (1792–1872), one of the three explorers (Blaxland, Lawson and Wentworth) who in 1813 were the first to successfully cross the Blue Mountains. The word 'falls' was included because of the high waterfalls in the locality.

Wentworthville *NSW,* a suburb of the City of Parramatta and the Municipality of Holroyd, recalls the Wentworth family which played a prominent part in the early history of New South Wales. D'Arcy Wentworth went to Norfolk Island in 1790 as assistant surgeon. On his return to Sydney in 1799 he was given a lease of land south of the site where the Town Hall stands today. His son, William Charles Wentworth, was the famous explorer of the Blue Mountains and one of Australia's leading political figures. D'Arcy Wentworth named his grant 'Wentworth Wood House' after an English property of that name.

Werribee *V* is a residential and rural area thirty-four kilometres south-west of Melbourne. Werribee Park estate was built in 1877 by wealthy landowners, the Chirnside brothers, and is a sixty-room Italianate mansion in spacious grounds beside the river. The original settlement was named Wyndham. The Aboriginal word *weariby* means 'backbone' or 'spine'. It is thought to have its origin in the tree-marked course of the Werribee River which stood out as a backbone on the flat and almost treeless plains surrounding the area.

West Lakes *SA* is a suburb west of the City of Woodville in the Adelaide metropolitan area. A company created the lake and suburban land areas by dredging the upper reaches of the Port River. The name was established by the West Lakes Development Act, 1969, an agreement between the South Australian Government and a private developer.

West Torrens *SA* is a corporate city close to the centre of the Adelaide metropolitan area. The river Torrens which flows through the city was named by Surveyor-General Colonel William Light, after Colonel Robert Torrens, Chairman of the South Australia Colonisation Commission, and father of

Sir Robert Richard Torrens, who became Premier of South Australia.

West Wyalong *NSW* probably comes from an Aboriginal word. The town of Wyalong, which was established in 1893 when gold was discovered, was named after an old sheep station. Wyalong Village was laid out in 1894 and a year later another township, previously known as Main Camp, was called West Wyalong.

Westbury *T* was first shown on Thomas Scott's Map of Tasmania dated 1830. The origin of the name is unknown but it could have been after the town in Wiltshire, England.

Western Australia As with the other states that separated from New South Wales, there was controversy over the name. The first governor, Sir James Stirling, wanted to call it Hesperia, 'land looking west' and in modern times Churchill has been proposed. But the colony has only ever been named Western Australia by British authorities. The western half of Australia was known as New Holland until British authorities commenced using the name Western Australia in 1828.

Western Port *V* is a bay lying south-east of Port Phillip Bay and adjoining Bass Strait. Discovered and named by George Bass on 5 January 1798, it was so-named because of its 'relative situation to every other known harbour on the coast'.

Westmead *NSW,* a suburb of the Municipality of Holroyd and the City of Parramatta in Sydney's western suburbs, was originally the west meadow of the Toongabbie Farm.

Weston *ACT* recalls the 'Weston' homestead that once stood in the Woden District.

Wetherill Park *NSW* is a suburb of Fairfield on the south-western outskirts of the Sydney metropolitan area. It is named after a Mr Wetherill who offered part of his property to the state government as a park around 1900. The offer was accepted and the park was named after the donor.

Whalan *NSW,* a locality in the City of Blacktown on the western outskirts of Sydney is named after James Whalan, who was granted land at Mount Druitt by Governor Darling in 1831.

Whale Beach *NSW* is located in Warringah Shire. Whale Beach is named after the Whale Beach Estate which was first subdivided in the early 1920s. This was named after a whale which was beached there at the turn of the century.

Wheeler Heights *NSW* is named after the Wheelers, one of the pioneering families of the Warringah Shire.

Wheelers Hill *V* is a locality in the City of Waverley in the Melbourne metropolitan area. Wheelers Hill was probably called after John Wheeler, an early settler, who settled beside this steep hill on the road to Fern Tree Gully.

White Cliffs *NSW* probably derives its name from the cliff-like outcrop of white, clayey limestone in which opals are found. Opal was first mined at White Cliffs in 1889. The township was proclaimed in 1898.

Whittlesea *V* is a shire located about twenty kilometres from central Melbourne. Whittlesea should have been spelt Whittlesey as it was named in 1853 by the surveyor of the township site, Robert Mason, after his birthplace Whittlesey Village in Cambridgeshire, England.

Whitsunday Passage *Q* was named by Captain Cook, who passed through it in the *Endeavour* on Whit Sunday (4 June) 1770.

Whittington *V* is a district east of Geelong. It is named after James Whittington who settled there in 1852 with his family and a number of other Isle of Wight families.

Whyalla *SA* is a city founded in 1913 on Spencer Gulf about 400 kilometres from Adelaide. The origin of its Aboriginal name is uncertain. Rodney Cockburn says it means 'a place of deep water', an appropriate name for what became an important iron ore port after 1915. The anthropologist Dr N.B. Tindale says that an Aboriginal tribe near Port Pirie had a word *wajala* or *wayalla* meaning 'west'. A tribe near Port Augusta had a word *waiala* meaning 'I don't know'. A hill one kilometre west of the township was already called Whyalla Hill before the town was surveyed and named.

Wickham *NSW,* a suburb of Newcastle, takes its name from a suburb of Newcastle upon Tyne. The name means 'the village by the creek'. It borders Throsby Creek.

Wide Bay *Q* was named by Captain Cook on 18 May 1770.

Wilberforce *NSW* was named by Macquarie in honour of William Wilberforce, the leader in the fight to abolish slavery.

Wilcannia *NSW* is an Aboriginal word meaning 'gap in the bank where the floodwaters escape'. The town developed in the 1860s as a result of the paddlesteamer traffic. At its height, it was the third largest river port in the country. Wilcannia lost its position as 'queen city' of the west in the

1920s after motor transport displaced the riverboats as the main method of transport.

Wiley Park *NSW* is a suburb of the Sydney municipality of Canterbury. It is named after Mr J.F. Wiley who bequeathed land for park and recreation purposes on certain conditions. Council was undecided as to accepting and a public meeting was called. Again, after much consideration, council decided to accept the offer but wanted a board of seven trustees. This resolution was rescinded in 1906 and the common seal was fixed to the conveyance of the land from Wiley's Estate to Council on 19 December 1906. The suburb was named after the park because the railway station was so named.

Williamstown *V* is an industrial city on the Gellibrand Peninsula which forms the south-eastern arm of Hobsons Bay. Williamstown was known to the Aborigines as Koort-boork-boork, meaning 'clump of she-oaks'. Governor Bourke inspected the Gellibrand Peninsula in 1837 and had it named William's Town after King William IV of England. Local government began in 1856 when Williamstown was created a borough. It became a city in 1919.

Willmot *NSW,* a locality in the City of Blacktown on the western outskirts of Sydney, is named after Thomas Willmot, (1851–1938) who was Blacktown Shire President from 1906–1910 and 1911–1914.

Willoughby *NSW* is a municipality on Sydney's lower North Shore. Willoughby comes from the name given to the parish by Surveyor-General Sir Thomas Mitchell who called it after James Willoughby Gordon (1773–1851), one of his quartermasters during the Peninsular War.

Wilmot *T* was named after Sir John Eardley-Wilmot, who was appointed Governor of Tasmania in 1843.

Wilpena Pound *SA* A mixture of Aboriginal and English words, Wilpena means 'place of bent fingers'. One theory is that it is so cold here at times that the Aboriginals were unable to bend their fingers, but a more likely explanation is that the area resembles a hand with the fingers, which represent the mountain peaks, bent upwards. Pound is used in the sense of a small enclosure such as those in which cattle are kept. The area was discovered by C.M. Bagot in 1850.

Wilson, Mount *see* **Mount Wilson**

Wilsons Promontory *V* was known to the Aboriginals as Wommoom. George Bass was the first European to see the promontory on 2 January 1798, during his whaleboat voyage from Port Jackson. Governor Hunter is believed to have named it after Matthew Flinders' friend, Thomas Wilson, of London. It was formerly known as Furneaux's Land after Captain Furneaux.

Wilsons Valley *NSW* is a locality on the main road to Perisher Valley. It is named after an old Scotsman who camped there for some time while prospecting for gold during the gold rush period some time before 1860. Wilson travelled to Jindabyne occasionally to purchase food and generally spent some time at the pub having a drop of Scotch, a few rums and a yarn with the boys before his fifteen kilometre journey home. He would always end by saying, 'Well, I'll have to go back to my valley'. The old hands passing the camp always remarked, 'There's old Wilson's valley'.

Wimmera District *V* is a large tract of pastoral country forming the whole of the north-western portion of Victoria, and covers an area of about 60 000 square kilometres. The area, which was a former roads district, consists mainly of vast, sandy and sparsely grassed plains, intersected with belts of Myall scrub and various forests. To the north-west are extensive swamps caused by the overflow of the Murray River. It is bounded in the east by the Avoca River and the west by the Victorian and South Australian border. The District was discovered by Major T.L. Mitchell in 1836. Its name came from an Aboriginal word meaning 'boomerang' or 'throwing stick'.

Windsor *NSW* was originally known as Green Hills. On 6 December 1810, Governor Macquarie christened all five towns now known as the Macquarie Towns. He renamed Green Hills, Windsor, because of its resemblance to Windsor in England.

Windsor *Q* is a locality in the Brisbane metropolitan area and is named after Windsor in England.

Windsor Gardens *SA* is a suburb in the corporate city of Enfield in the Adelaide metropolitan area. Windsor Gardens was named after Windsor in England. It was originally called only Windsor when it was advertised for sale by Nathaniel Hailes in October 1849. There is also a private town of Windsor north of Adelaide on the road to Port Wakefield.

Wingen *NSW* means 'fire', an appropriate name in view of the nearby Burning Mountain where a coal seam has been smouldering for centuries.

Wingen, Mount *see* **Burning Mountain**

Wingfield *SA* is a suburb in the city of Enfield in the Adelaide metropolitan area. Wingfield was named in May 1877 after

Richard William Wingfield, a private secretary to Governor Jervois.

Winmalee *NSW,* a locality in the Blue Mountains west of Sydney, was known as North Springwood until 1970 when the Geographical Names Board, in an endeavour to eliminate cardinal points of the compass from place names, suggested the name White Cross. The residents objected and on 28 April 1972, the name Winmalee was approved by the Minister for Lands.

Winston Hills *NSW* is a locality in the Baulkham Hills Shire on the north-western outskirts of Sydney. It was formerly Model Farm and Old Toongabbie. It was changed by land developers between 1960 and 1970 despite public opposition.

Winton *Q* was named by the first postmaster, Robert Allen, after his birthplace, a suburb of Bournemouth in England. Allen built a hotel-store in 1876.

Wisemans Ferry *NSW* is named after Solomon Wiseman (1777–1838) who arrived in Australia in 1806 as a convict sentenced for stealing twenty-four pounds worth of wood from a lighter on the River Thames. In 1811, he launched his ship, the *Hawkesbury Packet.* Another vessel, the *Hope,* was added soon afterwards. In 1812, Wiseman received a pardon and opened an inn in Bligh Street, Sydney. In 1817 both his vessels were wrecked, leaving him heavily in debt. He sold his remaining assets and took up a grant of land on the Hawkesbury where he built a residence. In 1821, he obtained a licence for an inn on the premises then known as The Sign of the Packet. In 1827 he established a ferry. As he grew older, Solomon Wiseman prospered. He acquired assigned convicts to work for him and was given a government contract to supply provisions.

Wittenoom Gorge *WA* was named after Frederick Francis Bividelt Wittenoom, a partner in Mulga Downs Station, by Mr George Hancock, father of Lang Hancock. In 1937 Lang Hancock began mining for long blue fibre asbestos in Wittenoom Gorge and did so until 1943, when the venture was taken over by C.S.R. A town was commenced in 1947 to provide homes and facilities for the mine workers. The town-site was gazetted on 9 March 1951 as Wittenoom Gorge and amended to Wittenoom on 5 April 1974. The asbestos mine was closed in 1966.

Wodonga *V* is a city on the Murray River, eleven kilometres west of Lake Hume. The area was named by Charles Huon in 1836 when he squatted there. When the township was surveyed in 1852, the name was changed to Belvoir after Lord Belvoir of Leicestershire, an English MP. The name reverted back to Wodonga in 1869. It is an Aboriginal word meaning 'edible nut'.

Wollombi *NSW* means 'meeting of the waters'.

Wollongong *NSW* was surveyed by Sir Thomas Mitchell, the Surveyor-General in 1834. He retained the Aboriginal name of Wollongong which is said to mean 'sound of the sea'.

Wollstonecraft *NSW,* a locality in North Sydney, is named after Edward Wollstonecraft (1783–1832) who obtained a grant of land on the North Shore in 1819.

Wonthaggi *V,* a township to the south-east of Western Port, comes from an Aboriginal word said to mean 'pull or drag along'.

Woodend *V* is the place where the track from Melbourne to the Castlemaine and Bendigo goldfields emerged from the Black Forest, a bushranger haunt in the 1850s, north of Mount Macedon.

Woodford *NSW* is a locality in the Blue Mountains west of Sydney. It was formerly known as Buss' Inn. It is understood that the name was changed to Woodford in 1874, this being the name of the mountain residence of A. Fairfax Esq. in that locality.

Woodford Bay *NSW,* a locality in the Municipality of Lane Cove, marks the boundary of a property named Woodford Park, owned by Rupert Kirk (1977–1850), a former captain in the Army Medical Corps, who had a grant of 130 hectares. Kirk used the land for the refining of oil and sugar and the manufacture of vinegar.

Woodforde *SA* is a suburb within the District Council area of East Torrens in the Adelaide metropolitan area. It was probably named by John Hallett, who held the first land grant over the original section, after his birthplace Woodford in Essex, England. Hallet arrived at Holdfast Bay in the barque, *Africaine,* well before the proclamation was read by the first governor in 1836. He brought with him a small flock of sheep. Hallet and his partner, Captain John Duff, were the first to export wool from South Australia when they consigned four bales to London in 1837. According to Rodney Cockburn's book, *What's In A Name,* Woodford was sometimes spelt with an "e" (Woodforde) because it was believed (incorrectly) that the area was named after Doctor John Woodforde, ship's surgeon with Colonel Light in the brig, *Rapid*. The name Woodforde was assigned under the Geographical Names Board Act in the Government Gazette dated 13 July 1978.

Woodridge *Q* was a name coined by Mr O. Stubbs, a saw-miller who subdivided land there in 1914. He called the subdivision 'Wood' because of the timber in the district and 'ridge' because it was the highest point on the rail line between Brisbane and Southport.

Woodroffe, Mount *see* **Mount Woodroffe**

Woodside *SA* was named after a village in Scotland by James Johnson of Mount Barker who purchased the township section in 1850. The plan of the town of Woodside dated 8 September 1856 was deposited in the General Registry Office as Plan 84 of 1857 signed by James and William Johnston, proprietors, and surveyed by E. Duval.

Woodville *SA* is a suburb in the corporate city of Woodville in the Adelaide metropolitan area. Woodville was named because the area was richly wooded, possibly with native pines because many areas in the region were marked 'the pinery'. The area was first subdivided by Emanuel and Judah Moss Solomon on behalf of Captain Thomas Lipson in 1849. John Bristow Hughes subdivided sections in 1855. He donated land and built St Margaret's Church at his own expense insisting that it should be named after his wife.

Woolamai, Cape *see* **Cape Woolamai**

Woollahra *NSW* is a municipality in the Sydney metro-politan area. Woollahra was the Aboriginal name for South Head. The white settlers called South Head 'the Lookout' because it was from this point that soldiers kept an anxious watch for the ships bringing much-needed supplies in the early days of the colony.

Woolloomooloo *NSW* is a locality in the City of Sydney. It is an old wharf-side suburb, and gets its name from a house of that name built by the first New South Wales commissary-general, John Palmer, in 1801. The name is derived from a local Aboriginal word related to kangaroos. It became a residential area when terraces were built there in the 1860s. The first wharf was also built there about this period.

Woolloongabba *Q* An Aboriginal term meaning 'whirling round'. As pronounced by the Aboriginals the accent was placed on the second syllable. Locally this Brisbane suburb is often contracted to The Gabba.

Woolooware *NSW,* in the Sutherland Shire, is an Aboriginal word meaning 'a muddy flat'. It was named by Surveyor Dixon in 1827.

Woolwich *NSW,* a locality on the shores of the Parramatta River, is named after the dockside district on the River Thames in South-East London.

Woomera *SA* An Aboriginal word meaning 'throwing stick'— an appropriate term for the site of the rocket range project.

Woronora *NSW,* in the Sutherland Shire, is an Aboriginal word meaning 'black rock'. It was originally Wooloonora. It is believed to have been named by Surveyor Dixon in 1828.

Woy Woy *NSW* is an Aboriginal word. It may mean 'deep water' or 'lagoon'.

Wrest Point *T* originally bore the name Gibbet Point because the gibbets were located there. Later it was known as Dunkleys Point after David Dunkley who had land there. George Robinson bought the point in 1898 and rebuilt the house there calling it 'St Helena'. Another owner bought the property about 1900 and chose the name 'Wrest Point'. The reason for the choice of word 'wrest', presumably meaning 'to seize by force', has remained a mystery. The name was transferred to the Wrest Point Riviera Hotel when it was developed just before World War II and is now applied to the casino which opened there in 1973.

Wungong *WA* is an area of Armadale, a town on the out-skirts of the Perth metropolitan area. Wungong Brook was previously named Marshall River by Surveyor Alfred Hillman in January 1835. The Aboriginal name of this stream was first recorded by Surveyor J.W. Gregory in 1844, and since that time Wungong (spelt in various ways) has been used on maps in preference to the name bestowed by Hillman. The town-site was proclaimed on 12 March 1909.

Wyee *NSW* means 'fire'.

Wyndham *WA* was probably named by Sir John Forrest after the son of Lady Barker, wife of the Governor, Sir Napier Broome. The town-site was declared on 1 September 1886. The name may also have been in honour of the Earl of Egremon, George O'Brien Wyndham who died in 1837.

Wynn Vale *SA* is named after Samuel Wynn, the wine maker, who owned land in the area.

Wynnum *Q* is a locality in the Brisbane metropolitan area and gets its name from an Aboriginal word meaning 'breadfruit tree'.

Wynyard *T* was named in the 1850s after E.B. Wynyard, a lieutenant-general of the New South Wales Corps.

Wyong *NSW* means 'place of running water'.

Yackandandah *V* is a mining town twenty-eight kilometres south of Wodonga. It is an Aboriginal word meaning 'hilly country'.

Yagoona *NSW* is a suburb of Bankstown, a city located in the western part of the Sydney metropolitan area. *Yagoona* is an Aboriginal expression for 'today'.

Yampi Sound *WA* comes from an Aboriginal word *yampee* meaning 'fresh water'. The name was given by J.L. Stokes in 1838 when he found fresh water there.

Yan Yean *V* is a locality of the Shire of Whittlesea on the outskirts of the Melbourne metropolitan area. *Yan-yean*, an Aboriginal word meaning 'boyish' or 'bachelor', was the name of an Aboriginal chief who signed John Batman's treaty.

Yanchep *WA*, a town in Wanneroo Shire north of Perth, is a corruption of an Aboriginal name Yanget meaning 'bullrushes', which grow wild along the edges of Lake Yanchep. The area was recorded by Captain George Grey in 1838 as Mau-bee-bee and Lake Yanchep was recorded by Surveyor Quinn in 1866.

Yaralla *NSW* is a locality in the Sydney municipality of Concord. It is an Aboriginal word meaning 'camp' or 'home'. It was the name given to the area by Thomas Walker.

Yarra River *V* rises near Mount Gregory in the Great Dividing Range and flows south-west to enter Port Phillip Bay. It was discovered in 1803 by the Surveyor-General of NSW, Charles Grime, who named it Freshwater River. Surveyor J.H. Wedge adopted the Aboriginal name Yarra Yarra, meaning 'ever flowing', for the river in 1835 while surveying the region for John Batman.

Yarralumla *ACT* is an Aboriginal place name first recorded on the 1829 map of Surveyor Robert Dixon.

Yarrambat *V* is a locality in the Shire of Diamond Valley on the outskirts of the Melbourne metropolitan area. Yarrambat was officially named in 1929. It had previously been known as Tanek's Corner and later as Hilton. *Yarrambat* is an Aboriginal word for 'high hills and pleasant views'.

Yarraville *V* is a suburb of the City of Footscray in the Melbourne metropolitan area. The name literally means 'town by the Yarra River'.

Yarrawarrah *NSW* in the Sutherland Shire is an Aboriginal word meaning 'place of echoes'.

Yarrawonga *V* is a township by Lake Mulwala on the Murray River. The name is an Aboriginal word meaning 'place where wonga pigeon nested'.

Yass *NSW* is a corruption of an Aboriginal word *yarh* meaning 'running water'. Hamilton Hume and William Hovell reached the area in 1824 and called the district McDougall's Plains.

Yatala *SA* is an area adjoining the northern boundary of the River Torrens in the Adelaide metropolitan area. Yatala was the Aboriginal name for the district. It means 'water running by the side of the river'. Yatala Vale is a suburb in the corporate City of Tea Tree Gully. According to the Australian Broadcasting Corporation, the correct pronunciation of this name is Yatter-luh, not YAHT-ah-lah.

Yea *V* is a township by the Goulburn River, forty kilometres south-east of Seymour. It is named after Colonel Lacey Yea who was killed at Sebastapol in 1855 during the Crimean War.

Yeerongpilly *Q* is a locality in the Brisbane metropolitan area and is an Aboriginal word meaning 'rain coming'.

Yennora *NSW*, a suburb of the City of Fairfield and the Municipality of Holroyd in Sydney's western suburbs, is an Aboriginal name meaning 'walking'. The New South Wales Railway Department built a siding platform in the area in 1927 and retained the name for the station.

Yeppoon *Q* is an Aboriginal word meaning 'thunder, roar or booming of the surf on the beach'.

Yirrigan *WA*, in Stirling City in the Perth metropolitan area, is an Aboriginal word meaning 'a high place'. It was chosen as a name for a subdivision in the State Housing Commission 'Mirrabooka Project Area' on 18 November 1954.

Yokine *WA* is in Stirling City in the Perth metropolitan area. For many years Mount Yokine was known as the location of a popular metropolitan golf club and the name was adopted when the land was subdivided for housing. The name Yokine Hill was originally applied to a survey point in

Williams Road in January 1922. The Aboriginal name means 'wild dog' and was chosen because of Native Dog Swamp close by.

York *WA* is Western Australia's oldest inland town. The first settlers arrived here in 1830 and Governor James Stirling announced that he had 'opened up the district for location under the name of Yorkshire which it was thought in some respects to resemble'. The name was contracted to York. Land was set aside for the York town-site in November 1830 and was gazetted in July 1835. Boundaries of the town-site were adopted by Executive Council and signed by the Governor on 19 April 1836. York is first shown on the plan of the South-West of WA in 1833, and is named after York in England. The name is presumed to have been suggested by two Yorkshire members of an exploration party of R. Dale's in October 1830.

York, Mount *see* **Mount York**

York Peninsula, Cape *see* **Cape York Peninsula**

Yorke Peninsula *SA* is a large promontory which separates Spencer Gulf and Gulf St Vincent. It was named by Captain Matthew Flinders, in 1802, after the Honourable Charles Philip Yorke, who was First Lord of the Admiralty. He later became Lord Hardwicke. He authorised the publication of Flinders Journal.

You Yangs *V* is a group of hills between Werribee and Geelong. Matthew Flinders climbed and named the main peak in 1803. You Yangs comes from the Aboriginal words *wurdi yowang* meaning 'big hill'.

Young *NSW* was originally known as Lambing Flat. It was renamed in 1861 in honour of the then governor of New South Wales, Sir John Young.

Yowie Bay (or Ewey [Ewie] Bay) *NSW* is the subject of two theories. The first is that it is a corruption of ewes (female sheep). The second is that yowie is an Aboriginal word meaning 'place of echoes'. Ewe is a Scottish name for a female sheep and *ewey* or *ewie* is a Scottish name for the ewe's lamb. *Yow* in Yorkshire means a female sheep (pronounced yo). *Yowie* is the *yow's* lamb in Yorkshire. In the Sutherland Shire, when the Honourable Thomas Holt was breeding sheep, he employed both Scottish and Yorkshire shepherds. The inlet (on the Hacking River) with its sheltered nooks where sheep had their lambs was accordingly known both as Ewie or Yowie Bay: where the lambs were born.

Yuleba *Q* comes from an Aboriginal word *ula* meaning 'blue waterlilies', and *ba* meaning 'place of'. There were several lagoons covered with large waterlilies near the home site of Old Yuleba, where the original township of Yuleba was located on the Moongool pastoral holding. It was proclaimed a town in 1872. The present town of Yuleba was originally called Baltinglass. It was proclaimed in 1878. With the opening of the western railway line to Baltinglass in 1879, Old Yuleba lost any reason for existence and people moved to the new railway town which became known as Yuleba.

Zeehan *T* was named in 1798 by Bass and Flinders after one of Abel Tasman's ships.

Zeil, Mount *see* **Mount Zeil**

Zetland *NSW,* a Sydney suburb, was named after the Earl of Zetland, a relative of the Governor, Sir Hercules Robinson. It was named in the 1870s when a continuation of Elizabeth Street was built to 'Zetland Lodge'.

Zillmere *Q* is a locality in the Brisbane metropolitan area and got its name from the Moravian missionary who was the original grantee.